Test Prep and Admissions

MCAT®/GRE® Biology Test Preparation Guide

for

Campbell • Reece

Biology

Seventh Edition

Test Prep and Admissions

MCAT®/GRE® Biology Test Preparation Guide

for

Campbell • Reece

Biology

Seventh Edition

San Francisco Boston New York
Cape Town Hong Kong London Madrid Mexico City
Montreal Munich Paris Singapore Sydney Tokyo Toronto

Editor-in-Chief: Beth Wilbur
Project Editors: Sarah Jensen, Megan Rundal
Editorial Assistant: Julia Khait
Managing Editor: Erin Gregg
Production Supervisor: Jane Brundage
Marketing Manager: Jeff Hester
Manufacturing Buyer: Stacy Wong
Production Services and Composition: TechBooks/GTS

Cover Photo: Linda Broadfoot

ISBN 0-8053-7178-8

Contents

MCAT Biology Practice Set

Passage I (Questions 1–6)

The development of the nervous system is characterized by an initial overproduction of neurons, followed by the subsequent elimination of excess nerve cells through cell death. This system appears to provide an efficient way of establishing proper nerve pathways through synaptic connections: a given synapse is more likely to form when many neurons, rather than only a few, grow toward a target neuron. Once the appropriate synapse is produced, the superfluous neurons degenerate.

Scientists have long been intrigued by the mechanisms that regulate the growth and elimination of these superfluous neurons. Experiments have shown that *nerve growth factor (NGF)* causes the axons of sympathetic neurons to grow in great abundance along chemical tracks left by the diffusion of NGF. NGF is one of several proteins known as neurotrophic agents that appear to promote neuron growth. Another trophic agent currently under study is *ciliary neurotroph factor (CNTF)*, which is believed to increase the growth of ciliary neurons and spinal motor neurons. The following assays were performed to test this hypothesis and the results are shown in Table 1 below.

Assay 1

In vivo treatment of developing chick embryos with CNTF*, in which test solution was injected into the embryos.

Table 1

Assay	% of neurons in which growth was induced	
	In vivo	*In vitro*
Ciliary neurons + CNTF	0	25
Ciliary neurons + dilution fluid	0	0
Spinal motor neurons + CNTF	30	30
Spinal motor neurons + dilution fluid	0	0

*Control assays were treated with an equal volume of the solution used to dilute CNTF.

Assay 2

In vitro treatment of separate cultures of embryonic ciliary neurons and embryonic spinal motor neurons with CNTF*, in which test solution was administered to the cultures.

1. From the information in the passage, it can be inferred that synapse formation:

    A. is a preprogrammed process with fully predictable results.
    B. is a completely random event.
    C. is a process that would occur much less efficiently without neurotrophic agents.
    D. is not absolutely essential for the formation of complex nervous pathways.

2. The experimental data shown in the table supports which of the following conclusions?

    A. *In vivo* growth of ciliary neurons is dependent on CNTF.
    B. Both *in vivo* and *in vitro* growth of ciliary neurons are dependent on CNTF.
    C. *In vivo* growth of spinal motor neurons is not dependent on CNTF.
    D. Neither *in vivo* nor *in vitro* growth of spinal motor neurons is dependent on dilution fluid.

3. According to the passage, NGF enhances the growth of sympathetic neurons. Based on this, which of the following physiological responses would most likely be compromised in an adult chicken if an NGF inhibitor were administered to the developing chick embryo?

    A. Inhibition of heart rate
    B. Stimulation of digestion
    C. Vasoconstriction of blood vessels in skeletal muscle
    D. Pupil dilation

4. Which of the following assays produced an un-expected result?

   A. *In vivo* treatment of ciliary neurons with CNTF
   B. *In vitro* treatment of ciliary neurons with CNTF
   C. *In vivo* treatment of spinal motor neurons with CNTF
   D. *In vitro* treatment of spinal motor neurons with CNTF

5. The process by which a developing neuron contacts a target neuron and forms a synapse is most analogous to which of the following biological processes?

   A. Migration of chromosomes during cell division
   B. Packaging and exocytosis of a secretory protein
   C. Binding of a hormone to its cell surface receptors
   D. Fertilization of an ovum by a spermatozoan

6. If the assays described in the passage were performed on adult chickens and no neuron growth was observed, which of the following would best explain the results?

   A. Ciliary and spinal motor neurons degenerate during maturation.
   B. Adult chickens can form new synapses.
   C. CNTF can induce neuron growth only during a critical period early in embryonic development.
   D. CNTF cannot induce neuron growth of embryonic ciliary neurons *in vivo*.

Passage II (Questions 7–11)

Several methods of gene therapy have been developed to insert foreign genes into cells with genetic defects. Microinjection of a gene into a target cell with a fine glass pipette has been successful in some cases, but is very time-consuming and requires a high level of expertise. Another approach is electroporation, in which DNA is stimulated to enter cells by exposure to electric shock; however, this procedure is traumatic to the cells. To date, the most effective method of introducing foreign genes into cells is through a viral vector. With this method, foreign genes enter the cell via a normal viral infection mechanism.

The genome of a virus may consist of DNA or RNA, and may be either single- or double-stranded. When certain DNA viruses infect a cell, their DNA is inserted into the host's genome. Once integrated, viral genes can be transcribed into mRNA, which is then translated into protein. By contrast, the genomes of simple RNA viruses are translated directly into mRNA by the enzyme RNA replicase; DNA never enters the process. In retroviruses, the RNA genome is transcribed by the enzyme reverse transcriptase into DNA, which is then inserted into the host's genome. The viral genes can then be expressed, directing the synthesis of viral RNA and proteins. Retroviruses consist of an outer protein envelope surrounding a protein core that contains viral RNA and reverse transcriptase. The retroviral RNA contains the three coding regions *gag*, *pol*, and *env*, which code for the viral core proteins, reverse transcriptase, and the coat protein, respectively.

So far, retroviruses seem a more promising tool for gene therapy than either DNA viruses or simple RNA viruses. A retrovirus containing a gene of interest enters a cell via receptor-mediated endocytosis, and its RNA is then transcribed into DNA. This DNA randomly integrates into the cellular DNA, forming a provirus that is copied along with the chromosomal DNA during cell division. Retroviral vectors are constructed such that the therapeutic gene takes the place of *gag*, *pol*, or *env*. Problems associated with this form of gene therapy include the possibility that random integration could lead to activation of an oncogene, the fact that integration can occur only in cells that can divide, and the limitation that gene expression cannot be precisely controlled due to the randomness of integration.

7. Simple RNA viruses make poor gene-therapy vectors because:

   A. their genomes are not large enough to accommodate a therapeutic gene.
   B. a therapeutic gene inserted in the form of RNA cannot be copied and passed on to daughter cells.
   C. they can infect only specific cell types.
   D. insertion of a therapeutic gene into an RNA genome renders it too unstable to be used as a vector.

8. *In vitro* experiments have shown that a retroviral delivery system for gene therapy is preferable to physical methods of delivering DNA into cells, such as microinjection or electroporation. Which of the following is the most likely explanation for this observation?

   A. Retroviral delivery permits greater control over the site of integration.
   B. Retroviral delivery produces a greater proportion of cells that successfully integrate the therapeutic gene.
   C. Retroviral delivery is less labor-intensive and less destructive to the cells.
   D. Retroviral delivery allows therapeutic genes to be inserted into all cell types.

9. Which of the following cells would NOT be good targets for gene therapy involving a retroviral vector?

   A. Liver cells, because they divide continually, so the effects of the therapeutic gene would be diluted.
   B. Skin cells, because they are continually sloughed off the surface of the skin, so the therapeutic gene would be lost.
   C. Mature neuronal cells, because they are not capable of division, so the therapeutic gene would not be able to integrate.
   D. Bone marrow cells, because they cannot be removed from the body.

10. Which of the following events must occur for a retrovirus carrying a therapeutic gene to successfully infect its target cell and integrate into the cell's genome?

    A. The target cell's surface receptors must bind the retroviral protein envelope.
    B. New virions must be produced.
    C. The retroviral genome must be translated by reverse transcriptase.
    D. The retroviral proteins encoded by *gag, pol,* and *env* must be synthesized by the vector virus after integration has occurred.

11. Once a therapeutic gene has been integrated into a target cell's DNA, the retroviral DNA will most likely:

    A. cause a nondisjunction event that corrects the genetic defect.
    B. be recognized as "foreign" by the host's immune system and degraded.
    C. replicate and form infectious virions.
    D. persist in the cell in a noninfectious form.

Questions 12 through 16 are NOT based on a descriptive passage.

12. The smallpox vaccine contains *vaccinia* virus, which has proven effective in providing active immunity to smallpox. The graph below plots the serum level of smallpox antibodies in a patient administered a smallpox vaccine on Days 1 and 50. The increase in serum level of smallpox antibody following the second vaccination can most likely be attributed to:

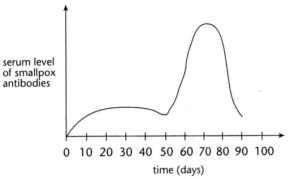

serum level of smallpox antibodies

0 10 20 30 40 50 60 70 80 90 100

time (days)

Figure 1

A. the smallpox infection caused by *vaccinia* virus.

B. the proliferation of B-lymphocytes exposed to *vaccinia* during the first vaccination.

C. the synthesis of smallpox antibodies by *vaccinia* virus and by the recipient's B-lymphocytes.

D. the fact that most vaccines require a minimum of 45 days for active immunity to be conferred on the vaccine recipient.

13. In patients with *myasthenia gravis*, a severe autoimmune disease of neuromuscular junctions, the body produces antibodies against the acetylcholine receptors of the muscle membrane (the sarcolemma), triggering their removal via phagocytosis. Based on this information, which of the following processes would be directly impaired by this condition?

A. Sarcomere shortening during muscle contraction

B. Calcium release by the sarcoplasmic reticulum

C. Action potential conduction across the sarcolemma

D. Acetylcholine synthesis

14. Which of the following factors would most likely cause an increase in left ventricle musculature (hypertrophy)?

A. High systemic blood pressure

B. Low systemic blood pressure

C. High pulmonary blood pressure

D. Low pulmonary blood pressure

15. What would be the most likely effect of infusing a concentrated solution of sodium chloride directly into the renal tubules of a healthy person?

A. Decreased urine volume, due to increased filtrate osmolarity

B. Increased urine volume, due to increased filtrate osmolarity

C. Decreased urine volume, due to decreased filtrate osmolarity

D. Increased urine volume, due to decreased filtrate osmolarity

16. A man with type AB blood marries a woman with type O blood. Which of the following are blood types that their chlildren might inherit?

A. Type A and type B

B. Type O and type AB

C. Type B and type O

D. Type A and type AB

Passage III (Questions 17–21)

The human menstrual cycle is regulated by a group of hormones that interact with one another via a complex system of positive and negative biofeedback mechanisms. The cyclic changes governed by these hormones result in the monthly preparation of the female reproductive system for fertilization and pregnancy. For example, prior to ovulation, the uterine endometrium proliferates and the uterine glands increase in size; following ovulation, the ovarian follicle atrophies into the corpus luteum, which maintains the endometrium through its secretion of estrogens and progesterone. If fertilization occurs, the embryo implants in the endometrium and the menstrual cycle ceases until at least six weeks after delivery, longer if the mother breastfeeds. If the ovum is not fertilized, menstruation occurs. During this process, the endometrium sloughs off and is expelled from the uterus along with the unfertilized ovum. Menstruation averages five days, and the entire cycle averages 28 days.

Plasma concentrations of four of the hormones that regulate the menstrual cycle—progesterone, 17 β-estradiol (one of the estrogens), luteinizing hormone (LH), and follicle-stimulating hormone (FSH)—were assayed in 100 female volunteers with 28-day menstrual cycles, each at an equivalent point in their cycle. The data for Assays A–G were averaged and appear below in Table 2.

Table 2

Hormones	Assay						
	A	B	C	D	E	F	G
	Day 1	Day 5	Day 10	Day 15	Day 19	Day 25	Day 27
Progesterone $\left(\dfrac{ng}{mL}\right)$	0.11	0.11	0.11	0.20	0.43	1.81	0.20
17-β-estradiol $\left(\dfrac{pg}{mL}\right)$	40.4	49.0	240.2	198.1	65.0	140.2	30.0
LH $\left(\dfrac{IRP - hMG}{mL}\right)$	8.5	10.5	20.3	50.2	14.0	10.0	8.0
FSH $\left(\dfrac{IRP - hMG}{mL}\right)$	10.0	12.0	10.0	23.1	8.0	7.0	8.0

Figure 2 represents the *in vivo* biosynthesis of progesterone from its precursor, pregnenalone:

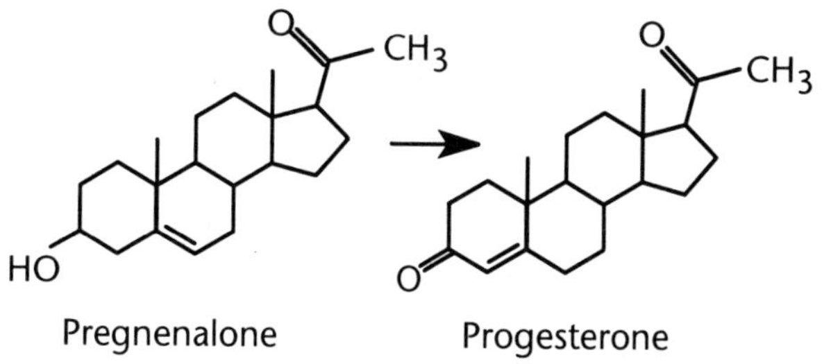

Pregnenalone   Progesterone
Figure 2

17. When 17 β-estradiol (a steroid hormone) contacts one of its target cells, it binds to an intracellular receptor and migrates to the nucleus; insulin (a peptide hormone) binds to extracellular receptors on the plasma membrane of its target cells. What is the most likely reason for this difference in mode of action?

    A. 17 β-estradiol is hydrophobic, and is therefore stable in cytoplasmic surroundings.

    B. 17 β-estradiol is lipid-soluble, and therefore easily traverses the plasma membrane.

    C. 17 β-estradiol is too small to bind to extracellular plasma membrane receptors.

    D. Insulin is too large to interact with DNA.

18. Based on the data in Table 2, what effect does the rising 17 β-estradiol plasma concentration appear to have on FSH secretion during the time interval between Assay B and Assay C?

    A. Positive biofeedback

    B. Competitive inhibition

    C. Negative biofeedback

    D. Allosteric activation

19. According to the passage, breast-feeding typically results *in secondary amenorrhea* (absence of menstrual periods). Given that breast-feeding is known to stimulate secretion of the hormone prolactin, which of the following explanations would best account for this effect?

    A. Prolactin inhibits ovulation, causing estrogen and progesterone secretion to decrease.
    B. Prolactin inhibits ovulation, causing estrogen and progesterone secretion to increase.
    C. Prolactin stimulates ovulation, causing estrogen and progesterone secretion to decrease.
    D. Prolactin stimulates ovulation, causing estrogen and progesterone secretion to increase.

20. Which of the following is the most likely explanation for the increase in progesterone and 17 β-estradiol plasma concentrations during the interval between Assay E and Assay F?

    A. The surge in LH secretion between Days 10 and 15 triggers the conversion of the ruptured ovarian follicle into the corpus luteum.
    B. 3-β-hydroxysteroid dehydrogenase, the enzyme that catalyzes the conversion of pregnenalone to progesterone, is inhibited by FSH.
    C. At Day 19, the uterine glands begin to secrete progesterone and 17 β-estradiol in anticipation of implantation.
    D. The surge in LH secretion between Days 15 and 19 increases 17 β-estradiol and progesterone production.

21. According to Figure 2, the biosynthesis of progesterone from pregnenalone results from which of the following?

    A. Reduction of a carbon-carbon single bond
    B. Oxidation of a carbon-carbon double bond
    C. Reduction of a hydroxy group
    D. Oxidation of hydroxy group

Passage IV (Questions 22–28)

DNA sequencing determines the precise base sequence of a given fragment of DNA. One such sequencing technique is the Maxam-Gilbert method, which separates DNA fragments that differ by only one nucleotide.

In the Maxam-Gilbert method, a sample containing multiple copies of a DNA fragment is radio-labeled with ^{32}P. The DNA sample is divided into four equal portions, and each is subjected to a different chemical treatment. One cleaves the fragment before guanine (G) only; another before either purine, adenine (A) or guanine; a third before cytosine (C) only; and a fourth before either pyrimidine, thymine (T) or cytosine. If carried to completion, these reactions would cleave the fragment every time its particular base(s) appeared. However, limiting the reagent concentration and the duration of the reactions allows each fragment to be cleaved only once (on average). The procedure for cleaving guanines, for instance, produces a series of fragments of different lengths, each ending before a different guanine in the original molecule. Thus, all of the positions where a particular residue is found are represented in the series of fragments. The unlabeled fragments are ignored.

The products of the reactions are separated and their lengths identified by standard biochemical means. The four reaction mixtures are loaded onto four lanes of a polyacrylamide gel and subjected to electrophoresis. Larger residues migrate more slowly downward through the gel and the fragments spread out in a series of invisible bands. The positions of the fragments are then visualized using autoradiography: when the gel is held up to a photographic film, the radioactive phosphates expose the film, and the bands appear on the developed negative. DNA sequences up to 200 bases long can be analyzed in one set of experiments. Since molecular DNA is composed of two complementary strands, the first base in any chain can be identified by sequencing and identifying the final base of its complement.

Figure 3 shows the autoradiograph produced from Maxam-Gilbert sequencing of a short fragment of DNA. The band that appears at the top of the gel in all four lanes is the unreacted DNA fragment.

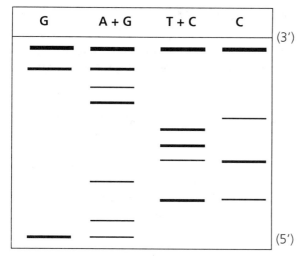

Figure 3

22. Why is the DNA fragment radio-labeled with ^{32}P at the beginning of the Maxam-Gilbert sequencing procedure?

    A. To cleave the fragment before specific bases
    B. To divide the sample into four equal portions
    C. To limit the duration of the cleavage reactions
    D. To identify the polarity of the DNA fragment

23. If a fragment of DNA was found to have the sequence 5'-AGACCTATG-3', which of the following RNA fragments would form a DNA-RNA hybrid with it?

    A. 5'-AGACCTATG-3'
    B. 5'-CATAGGTCT-3'
    C. 5'-CAUAGGUCU-3'
    D. 5'-UCUGGAUAC-3'

24. After applying the Maxam-Gilbert method to a DNA fragment of unknown sequence, a researcher would identify a base as a guanine if a band appeared in:

    A. the G lane only.
    B. the A + G lane only.
    C. both the G lane and the A + G lane.
    D. neither the G lane nor the A + G lane.

25. What is the base sequence of the DNA fragment depicted in Figure 3?

    A. 5'-GGACTTCAGA-3'
    B. 5'-GACACTTCAAG-3'
    C. 5'-GAACTCACAG-3'
    D. 5'-GATATCTAAG-3'

26. Which of the following is NOT a basic premise underlying the Maxam-Gilbert sequencing method?

    A. Each of the four chemical treatments cleaves the DNA fragment before only one nitrogen base.
    B. DNA contains the nitrogen bases adenine, guanine, thymine, and cytosine.
    C. The lengths of small DNA fragments can be compared by standard biochemical means.
    D. A single chemical treatment can produce labeled fragments of varying lengths.

27. According to the information in the passage, if the adenine + guanine cleavage reaction were allowed to continue to completion, how would this affect the appearance of the developed autoradiogram of the gel electrophoresis?

    A. A greater number of distinct bands would appear in the A + G lane.
    B. A smaller number of distinct bands would appear in the A + G lane.
    C. A smaller number of distinct bands would appear in the A lane.
    D. The number of unlabeled fragments in the A + G lane would decrease.

28. If the ^{32}P-radio-labeled DNA fragment 5'-(^{32}P)ACTATG-3' were subjected to the Maxam-Gilbert method, which of the following fragments would produce visible bands in either the C lane or the T + C lane?

    I. (^{32}P)A
    II. (^{32}P)AC
    III. (^{32}P)ACT
    IV. (^{32}P)ACTA
    V. (^{32}P)ACTAT

    A. I and II only
    B. II and IV only
    C. I, II, and IV only
    D. I, II, III, and IV only

Questions 29 through 30 are NOT based on a descriptive passage.

29. Which of the following processes is NOT ATP-dependent?

    A. Exocytosis of synaptic vesicles at a nerve terminal
    B. Movement of urea across a cell membrane
    C. Movement of Ca^{2+} from the sarcoplasm into the sarcoplasmic reticulum
    D. Export of Na^+ from a neuron

30. In Figure 4 below, the reaction coordinate shows the progress of an enzyme-catalyzed reaction. If an inhibitor of the enzyme were added to the reaction vessel, how would the reaction most likely be affected?

    A. The free energy of the reactants would decrease.
    B. The free energy of the products would increase.
    C. The activation energy would increase.
    D. There would be no change.

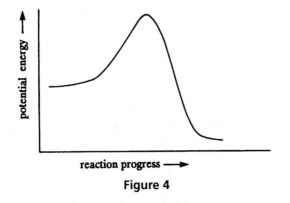

Figure 4

Passage V (Questions 31–35)

A researcher performed the following set of experiments in order to investigate the metabolism of two different strains of bacteria, Strain 1 and Strain 2.

Experiment 1

Strains 1 and 2 were incubated in separate broth cultures for 24 hours at 37°C. A sample of each culture was streaked onto three different plates—A, B, and C—each containing a different starch-agar medium; the plates were then incubated for another 48 hours at 37°C. The plates were then examined for surface colony growth and stained with iodine solution to determine the extent of starch digestion.

Table 3

	Surface colony growth			Starch digestion		
	A	B	C	A	B	C
Strain 1	+	+	+	−	−	−
Strain 2	+	+	−	+	+	−

Key: + = growth; − = no growth

Experiment 2

The two strains were incubated in the same manner as in Experiment 1. Two 100-mL portions of agar were poured into two beakers, which were maintained at 43°C. Next, 0.2 mL of broth culture from Strain 1 was pipetted into the first beaker, and 0.2 mL of broth culture from Strain 2 was pipetted into the second beaker. The agar was swirled around to distribute the bacteria evenly through the media, and then poured onto plates. These plates were incubated for 48 hours at 37°C and then examined for colony growth both on the agar surface and lower down within the oxygen-poor agar layer.

Table 4

	Surface colony growth	Deep-agar colony growth
Strain 1	+	−
Strain 2	+	+

Key: + = growth; − = no growth

31. All of the following structures are found in both Strain 1 bacteria and Strain 2 bacteria EXCEPT:

    A. circular DNA.
    B. mitochondria.
    C. cell walls.
    D. ribosomes.

32. Based on the results of Experiment 2, Strain 2 bacteria would most likely be classified as:

    A. obligate anaerobes.
    B. obligate aerobes.
    C. facultative anaerobes.
    D. chemoautotrophs.

33. Which of the following statements best accounts for the experimental observations obtained for Strain 1 bacteria in Experiment 1?

    A. Strain 1 bacteria do not possess the enzymes necessary to digest starch.
    B. Strain 1 bacteria do not use starch for their first 48 hours of growth.
    C. Strain 1 bacteria grow best under oxygenated conditions.
    D. Strain 1 bacteria cannot grow on starch-agar medium C.

34. The researcher decided to incubate the plates from Experiment 1 for three more days; the results obtained were identical to those from Experiment 1 EXCEPT that starch digestion was observed for Strain 1 on all three starch-agar media. Which of the following conclusions might be drawn from these observations?

    I. Strain 1 requires longer incubation times to digest starch.
    II. Strain 2 requires oxygen for its early stages of growth.
    III. Strain 2 cannot digest the starch in starch-agar medium C.

    A. I only
    B. III only
    C. I and III only
    D. I, II, and III

35. Suppose that the researcher repeats Experiments 1 and 2, and this time, starch digestion is observed for Strain 2 in all of the starch media. These results can most likely be explained by the occurrence of which of the following processes?

    A. Mutation
    B. Transduction
    C. Nondisjunction
    D. Meiosis

Questlons 36 through 38 are NOT based on a descriptive passage.

36. Animal studies have shown that certain lesions to the mesodermal embryonic primary germ layer may simulate the development of a rare human condition known as spina bifida, a congenital fissure in the lower vertebrae. Besides the spinal column, what other structures would most likely be affected by such lesions?

    I. Muscles
    II. Blood vessels
    III. Skin and hair
    IV. Intestinal epithelium

    A. I only
    B. I and II only
    C. II and III only
    D. III and IV only

37. Curve C in Figure 5 below represents the normal hemoglobin dissociation curve at 38° C and physiological pH (7.4). Which curve most likely corresponds to the hemoglobin dissociation curve for a patient suffering from acidosis (low blood pH)?

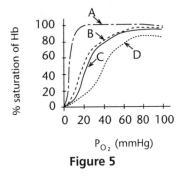

Figure 5

    A. Curve A
    B. Curve B
    C. Curve C
    D. Curve D

38. Which of the following graphs best represents the optimal pH for pepsin (a protease) activity?

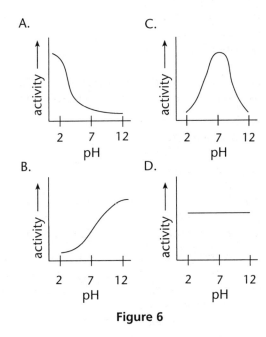

Figure 6

MCAT Biology Practice Set: Answers and Explanations

ANSWER KEY

1. C (See Concept 48.4)
2. D
3. D (See Concept 48.5)
4. A
5. D (See Concepts 48.4 and 48.7)
6. C (See Concept 48.7)
7. B (See Concept 20.5, pp. 403–404)
8. C (See Concept 20.5, pp. 403–404)
9. C (See Concept 20.5, pp. 403–404)
10. A (See Concept 20.5, pp. 403–404)
11. D (See Concept 20.5, pp. 403–404)
12. B (See Concept 43.3, p. 914)
13. C (See Concept 43.5, p. 917; Concept 48.4, p. 1024)
14. A (See Concept 42.3, pp. 876–877)
15. B (See Concept 44.5)
16. A (See Concept 14.3, p. 262)
17. B (See Concept 11.1, p. 203)
18. C (See Concept 46.4, pp. 973–977)
19. A (See Concept 46.4, pp. 973–977)
20. A (See Concept 46.4, pp. 973–977)
21. D (See Concept 4.3, p. 64)
22. D
23. C (See Concept 17.2)
24. C (See Concept 20.2, p. 393)
25. B
26. A
27. B (See Concept 20.2, p. 393)
28. C (See Concept 20.2, p. 393)
29. B (See Concept 8.3)
30. C (See Concept 8.4)
31. B (See Concept 6.2, pp. 98–99)
32. C (See Concept 27.2)
33. B
34. C
35. A (See Concept 27.1, pp. 537–538)
36. B (See Concept 47.1, pp. 999–1001)
37. D (See Concept 42.7, p. 892)
38. A (See Concept 41.4, pp. 857–858)

EXPLANATIONS

Passage I (Questions 1–6)

1. C The main concept in this passage is that the formation of synaptic connections in the nervous system is a complex procedure, somewhat comparable to shooting in the dark. Since neurons have no way of sensing each other, the probability that a single neuron growing toward a target neuron will form a synaptic connection is quite low. To compensate for this, the developing nervous system has evolved the following mechanism: it initially produces an abundance of neurons that grow toward a single target neuron. Once the proper synapse has formed, the superfluous neurons degenerate. Compare this to someone trying to shoot an arrow at a bulls-eye. The chances that one arrow will hit the center are much greater if many arrows are shot. The arrows that don't hit the target merely fall by the wayside. This analogy is appropriate because, as you're told in the passage, the growth of the neurons toward the target is not completely random; it is aimed, to some degree, by the neurotrophic agents that control neuron growth. This implies that without neurotrophic agents, synapse formation would occur much less efficiently; thus, choice (C) is right. Choice (A) is incorrect because, as just explained, the process of synapse formation certainly does NOT have predictable results. Choice (B) is incorrect because we also know that the process of synapse formation isn't completely random; it's under the influence of neurotrophic agents. Choice (D) is incorrect because the passage states that proper nerve pathways are established through synaptic connections. This means that synapse formation IS absolutely essential for the formation of complex nervous pathways.

2. D The assays were performed in order to determine whether CNTF increases the growth of ciliary neurons and spinal motor neurons both in vivo and in vitro. As a control, both types of neurons were also treated with the fluid used to dilute CNTF in the experiment—again, under both in vivo and in vitro conditions. The dilution fluid was not expected to increase neuron growth; that's why it was used as the control. Anytime CNTF promoted neuron growth, it provided support for the hypothesis behind the assays. Turning now to the answer choices, since growth was not induced in vivo ciliary neurons treated with CNTF, then choice (A) cannot be concluded, and must therefore be wrong. And although it might be concluded that in vitro growth of ciliary neurons is dependent on CNTF, since 25% of the neurons in that assay experienced growth, as just discussed this is NOT true for in vivo ciliary neurons. Therefore, choice (B) is also incorrect. Since 30% of spinal motor neurons treated in vivo with CNTF experienced growth, one would conclude that this growth was dependent on CNTF; thus choice (C) is also wrong. Choice (D), however, is supported by the data because NONE of the control assays were expected to cause neuron growth, and none of them did. Therefore, it can be concluded that neither in vivo nor in vitro growth of spinal motor neurons, or for that matter of ciliary neurons, is dependent on the dilution fluid used in the preparation of CNTF.

3. D   This question takes a tiny little piece of information from the second paragraph and uses it as a springboard. To answer it, you have to know the differences between the parasympathetic and sympathetic divisions of the autonomic nervous system. These two divisions often elicit antagonistic responses when they innervate the same organ. In general, the parasympathetic division elicits physiological responses that conserve and gain energy, such as slowing down heart rate, vasoconstriction blood vessels in skeletal muscle, constricting pupils, and decreasing metabolic rate; the sympathetic division, on the other hand, stimulates responses that expend energy and prepare an organism for action, such as inhibiting digestive processes, increasing heart rate, vasodilating blood vessels in skeletal muscle, dilating pupils, and increasing metabolic rate. These responses are commonly known as the "flight-or-fight" responses. Based on this, it can be inferred that if a developing chick embryo were given an NGF inhibitor, then the axons of sympathetic neurons would not grow in abundance, and the proper synaptic connections would not be formed. And this means that some of the nervous pathways leading to the physiological responses associated with sympathetic innervation might very well be compromised in an adult chicken. Choice (D), pupil dilation, is the only sympathetic response among the four choices; therefore, it's the only one that might be compromised, and is the correct answer.

4. A   In this experiment, it was expected that anywhere that ciliary neurotrophic growth factor was introduced, neuron growth would be induced. Thus, it's logical that anywhere CNTF was omitted, as in the four control experiments with dilution fluid, growth should not have been induced. However, in the in vivo assay of ciliary neurons plus CNTF, you notice that even though growth factor was introduced, no neuron growth occurred. This is an unexpected result, especially since growth was induced both in vivo and in vitro for the spinal motor neurons, and in vitro for the ciliary neurons. Therefore, choice (A) is the correct answer. There are many possible explanations for this result; one is that the experimental conditions may simply have gone awry. Another is that perhaps CNTF simply does not work this way within the body—that is, CNTF's in vitro neurotrophic effects may not be due to the same mechanism that normally causes ciliary neurons to grow in vivo, so that even though it may serve as a neurotrophic agent for embryonic ciliary neurons in vitro, it doesn't in vivo. None of the remaining answer choices are correct, because they're all expected results.

5. D   The process by which a target neuron forms a synapse with a developing neuron is largely determined by the probability that a neuron growing toward a target neuron actually synapses with it. The chances of this occurring would be quite low were it not for neurotrophic agents, which increase the number of neurons growing toward a single target. So during development, a large number of neurons are overproduced so as to ensure that at least one of them forms the proper synapse. Once the synapse has formed, the remaining neurons, which are no longer needed, will degenerate. Of the four answer choices, this process is most analogous to the fertilization of an ovum by a spermatozoan.

Sperm swimming through the female reproductive tract toward the egg are somewhat directed by chemical signals, just as neurotrophic agents direct neurons in their growth. Yet there is still the possibility that the ovum won't be fertilized, just as there's a chance that a single developing neuron won't synapse with its target neuron. To increase this probability, nature has dictated that millions of spermatozoa simultaneously attempt to fertilize the same egg, much like the initial overproduction of neurons increases the probability that the proper synapse will form. Thus, choice (D) is correct. Choice (A) is incorrect because there is no question of probability in the

KAPLAN
Test Prep and Admissions

migration of chromosomes during cell division. This process is precisely controlled; the chromosomes are guided toward their objective—the cell poles—by spindle fibers that are attached to their centrioles. Choice (B) is incorrect because the packaging and exocytosis of a secretory protein is not in any way similar to the development of a neural pathway. It's not as if a thousand molecules of the same protein are synthesized in the hopes that one of them makes it into a secretory vesicle and out of the cell. Finally, choice (C) is also incorrect because hormones are released into the circulatory system in minute quantities, and then travel to their target cells, where they bind with surface receptors specific for them. Hormones are not produced in excess, nor is there only a single cell with a single receptor that is receptive to the hormones' effects.

6. C According to information in the question stem, when the same exact procedures that had been performed on ciliary neurons and spinal neurons of embryonic chicks were performed on adult chickens instead, CNTF failed to induce any growth. The only difference between this set of experiments and the ones described in the passage is the age of the chickens. All four answer choices deal with this age discrepancy, so it makes sense to evaluate each of them to determine which best explains the experimental results. Choice (A) says that ciliary and spinal motor neurons degenerate during maturation. Not only is this false, but there's absolutely no evidence for it in the passage. What the passage does tell you is that the superfluous neurons produced by the developing embryonic nervous system degenerate after a synapse has properly formed. Thus, choice (A) is incorrect. If choice (B), which says that adult chickens can form new synapses, were actually true, then you would expect that CNTF would have induced neuron growth in the chickens. Since this did not occur, and since it's not clear from the passage whether adult chickens can or can't form new synapses, choice (B) must also be wrong. Choice (C), however, does provide a plausible explanation for the lack of CNTF-induced neuron growth in the adult chickens: neuron growth can only be induced during a critical period of embryonic development. In other words, there is a limited period of time during which CNTF can induce the growth of ciliary and spinal motor neurons, and this period occurs during embryonic growth. After this critical period has passed, ciliary and spinal motor neurons are no longer sensitive to

the effects of CNTF. So, choice (C) is correct. As for choice (D): While the claim that CNTF can't induce growth of embryonic ciliary neurons in vivo IS supported by the results of the experiments described in the passage, this does not account for the lack of ciliary neuron growth and spinal motor neuron growth in adult chickens treated with CNTF; therefore, choice (D) is wrong.

Passage II (Questions 7–11)

7. B The key to this question is the fact that whereas simple RNA viruses are poor vectors for gene therapy, retroviruses are good ones. So the factor that makes simple RNA viruses bad vectors must be something that is not true of retroviruses. The major difference between the two is that retroviral genomes get reverse transcribed into DNA. Remember, the passage tells you that the genome of a simple RNA virus is directly transcribed into messenger RNA. If a simple RNA virus were used as the vector for a therapeutic gene, then the gene would not be replicated and passed on to daughter cells, because an RNA gene cannot become integrated into the cell's DNA genome. Integration is a necessary step of gene therapy and can only occur if the gene is in the form of DNA. Since integration cannot occur, the viral RNA is rapidly degraded by the cell. Choice (A) is wrong because there is no evidence in the passage that RNA viruses have smaller genomes than either DNA viruses of retroviruses, or that they might therefore not be able to accommodate a therapeutic gene. Choice (C) is also incorrect, because although it's true that certain viruses can infect only specific cell types, this is true for all viruses, not just simple RNA viruses, so it would not be a factor that would distinguish simple RNA viruses in particular. And, in any case, the problem of cell type specificity on the part of the virus is not even discussed in the passage, so there's no evidence to support this choice. Choice (D) is incorrect since there is also no evidence in the passage to support the statement that incorporation of a therapeutic gene into an RNA genome would render it too unstable to be used as a vector. Instead, the passage implies that it is possible to produce vectors carrying therapeutic genes from all kinds of viruses, but that retroviruses make the best vectors for some other reason—which is the reason described by correct choice (B).

8. C The passage described microinjection as being a time-consuming technique requiring a high level of expertise, an electroporation as a technique that is traumatic for the cells. Neither of these criticisms is applied to retroviral delivery, which is described instead as occurring via a "normal" infection mechanism. Normal retroviral infection does not require any technique more complicated than adding the virus to the cells to be infected, and it does not harm the cells during the actual infection process, since the virus enters a lysogenic cycle. Thus, choice (C) is correct. Choice (A), which states that the site of gene integration can be more precisely controlled using a retroviral delivery system than by using physical methods, is wrong because the passage states that the retroviral DNA integrates into the cellular DNA at a random location—that is, not precisely at all! Choice (B) is also incorrect, since there is no evidence in the passage to suggest that a retroviral delivery system allows a greater proportion of its target cells to integrate the therapeutic gene than does electroporation or microinjection. And although you are not expected to know this, microinjection is probably the most efficient method of gene therapy in terms of the proportion of treated cells that are infected with the therapeutic gene. Finally, choice (D) is a wrong answer since the passage states that provirus integration occurs during RNA replication; thus, only cells that can divide are amenable to retroviral gene therapy.

9. C As you're told in the passage, a retrovirus integrates its DNA into cellular DNA during DNA replication, mainly during cell division, so it can only integrate into the chromosomes of cells that are capable of dividing. Of the four cell types listed as answer choices, only neuronal cells cannot divide. Choices (A) and (B) are both wrong, and for similar reasons. Liver cells would actually be GOOD targets for retroviral gene therapy because they divide continuously. Continuous division means that the therapeutic gene would be replicated along with the cellular DNA and inherited by the daughter cells, thus ensuring a continuous supply of the therapeutic gene product; thus, choice (A) is wrong. Likewise, though dead skin cells are continually sloughed from the surface of the skin, they are replaced by living ones through cell division. The retrovirus could only successfully infect live cells, and once it had infected these cells, the therapeutic gene would not be lost, but rather inherited by the daughter skin cells. Choice (D) is incorrect because bone marrow cells

are good targets for gene therapy involving a retroviral vector, for the very reason that they can be successfully removed from the body, infected, and then replaced. Moreover, bone marrow cells do divide and eventually produce important blood cell types, which makes them particularly fruitful subjects for gene replacement therapy. As you may be aware, bone marrow replacement has been used in recent years to treat lymphomas.

10. A For a retrovirus carrying a therapeutic gene to successfully infect and integrate into its target cell's genome, several events must occur. The first stage of infection is the binding of the retrovirus' protein envelope to receptors on the surface of the target cell. This facilitates the entry of the retrovirus into the cell, which consists of RNA. The RNA genome is first transcribed into DNA by the enzyme reverse transcriptase. You might therefore have been tricked by choice (C), which actually says that the retroviral genome must be translated by reverse transcriptase; however, translation is a different process. The stage of protein synthesis during which a strand of messenger RNA transcribed from the genome is used to produce a strand of amino acids is translation. Thus, choice (C) is also wrong. Choice (D) would not occur during successful gene therapy; the retroviral proteins encoded by the genes gag, pol, and env are not synthesized after integration has already occurred. Gag and env, which code for the retroviral core proteins and protein envelope, respectively, would be synthesized after infection only if the retrovirus entered a lytic cycle. And although the pol region codes for reverse transcriptase, which is necessary for integration, reverse transcriptase is synthesized before integration occurs; so choice (D) is incorrect.

11. D This one can easily be answered by the process of elimination. Choice (A) does not make sense. Nondisjunction is either the failure of sister chromatids to properly separate during mitosis, or the failure of homologous chromosomes to properly separate during meiosis—the net result being that some daughter cells inherit multiple copies of one chromosome, while others lack the chromosome entirely. In any case, nondisjunction is not at all desirable, and is in fact typically lethal. Thus, choice (A) is wrong. Choice (B) is wrong because the host's immune system would have no opportunity to come into contact with and recognize the retroviral DNA as foreign since the viral DNA integrated along with the therapeutic gene into the host cell's DNA. Choice (C) is wrong because in order for replication of the retroviral DNA and formation of infectious virions to occur, the virus would have had to enter a lytic cycle, which does not occur in successful gene therapy. Remember, if the retrovirus was lytic, then the host cell would be killed, and this is clearly not the goal of retroviral gene therapy. The goal is to introduce a healthy gene into a genetically defective cell and produce its protein product without causing an infection, which necessarily implies that the retroviral DNA persists in the host DNA in an noninfectious form.

Discrete Questions

12. B Vaccines consist of attenuated—that is, weakened or inactive—bacterial or viral forms. Vaccines are specifically designed to "fool" the body into synthesizing antigens against a particular pathogen, without actually causing the disease typically associated with that pathogen. Therefore, choice (A) is wrong, because a smallpox vaccine containing vaccinia virus does not cause smallpox. The reason vaccinia is used in the smallpox vaccine is that its protein coat contains antigens similar enough to those found on the smallpox virus that they stimulate the proliferation of B lymphocytes that will then produce antibodies specific for both vaccinia virus and smallpox virus. The vaccine recipient is thus protected against a future smallpox infection. This type of acquired immunity is known as active immunity. Active immunity has two phases: first, the B lymphocytes differentiate into either plasma cells or memory cells. The plasma cells immediately start to synthesize antibodies; the memory cells remain inactive, but retain surface receptors specific for the vaccine's antigens. This is known as the primary response. Upon subsequent exposure to the same antigen, such as a second vaccination or an exposure to the infectious agent, these same memory cells elicit a greater and more immediate proliferation of B lymphocytes; and this is known as the secondary response. On the graph, the curve following the first smallpox vaccination corresponds to the primary response, and the curve following the second smallpox vaccination corresponds to the secondary response. Thus, choice (B) is correct. As for choice (C), although the increase in the serum level of smallpox antibody can be attributed to the synthesis of smallpox antibodies, they are synthesized by the recipient's B lymphocytes, not by vaccinia

virus. Being a virus, vaccinia can't synthesize anything but the proteins and nucleic acid it needs to replicate itself, and it can only do that with the use of a host cell's genetic machinery; it certainly can't synthesize antibodies, which are only produced by multicellular organisms. And there's no evidence in the question stem to support the claim that the smallpox vaccine or any other vaccine requires a minimum of 45 days to confer active immunity, choice (D).

13. C A neuromuscular junction is composed of the presynaptic membrane of a neuron and the postsynaptic membrane of a muscle fiber (which is called the sarcolemma). The two are separated by a synapse. In response to an incoming action potential, the presynaptic membrane of a neuromuscular junction releases the neurotransmitter acetylcholine into the synapse. Acetylcholine diffuses across the synapse and binds to acetylcholine receptors on the sarcolemma, causing the membrane to depolarize and generating an action potential. The action potential causes the sarcoplasmic reticulum to release large amounts of calcium ions into the sarcoplasm, which is the cytoplasm of a muscle fiber. This in turn leads to the shortening of the muscle fiber's sarcomeres, which contracts the muscle. According to the question stem, myasthenia gravis causes the body to produce antibodies that trigger the removal of the acetylcholine receptors found on muscle fibers. Without these receptors, acetylcholine can't bind to the membrane and trigger its depolarization. And if this doesn't occur, then neither can any of the events that normally follow it. So, of the four choices, the one that would be directly impaired by myasthenia gravis would be the conduction of an action potential across the sarcolemma: therefore, choice (C) is correct. Choices (A) and (B) are wrong, because although they too are impaired by myasthenia gravis, they're dependent on the conduction of an action potential, so it can't really be said that they are directly impaired. Choice (D), acetylcholine synthesis, is wrong because it's not at all affected by myasthenia gravis; acetylcholine is synthesized by the neuron itself, and this process is independent of whether or not acetylcholine receptors are present on the postsynaptic membrane.

14. A Muscular hypertrophy is an increase in muscle mass, which is due to an enlargement of its constituent cells. The most likely cause of hypertrophy in any muscle is an increased workload. Therefore, the most likely cause of hypertrophy of the left ventricle would be an increase in workload for the left ventricle. The left ventricle is responsible for pumping blood into the aorta and supplying the force necessary to propel it through the rest of the systemic circulation. Therefore, an increase in systemic blood pressure would increase the resistance of systemic circulation and force the left ventricle to work harder to overcome it; and as a result, the left ventricle would gradually hypertrophy. Thus, choice (A) is right and choice (B) is wrong. Choices (C) and (D) are wrong because pulmonary circulation is propelled by the RIGHT ventricle, and so any changes in pulmonary blood pressure would primarily affect the right ventricle.

15. B Water reabsorption in the kidneys is directly proportional to the osmolarity of the interstitial tissue, relative to the osmolarity of the filtrate. When the osmolarity of the filtrate is lower than the osmolarity of the kidney tissue, the tendency is for water to diffuse out of the nephron; when the osmolarity of the filtrate is higher, the tendency is for water to diffuse into the nephron. Infusing the nephron of a healthy person with a concentrated sodium chloride solution increases the filtrate osmolarity; thus, water will diffuse into the nephron to try and compensate for this change. So, you can rule out choices C and D. And since the volume of urine excreted is inversely proportional to the amount of water reabsorption, there will be a concomitant increase in urine volume. Therefore, choice (A) is wrong and choice (B) is correct.

16. A To answer this question correctly, you need to understand the genetics of the ABC blood system; that is, you need to know that the A and B alleles are codominant to the O allele. Thus, a person with the genotype AA or AO has a type A blood; a person with the genotype BB or BO has type B blood; a person with the genotype OO has type O blood. A man with type AB blood has the genotype AB and can therefore produce with either the A allele or with the B allele. A woman with type O blood has two O alleles and therefore only produce gametes of the O allele. So if this couple has children, the only two genotypes their children can possibly inherit with respect to the blood groups are AO and BO, which corresponds to the phenotypes type A blood and type B blood, respectively.

Passage III (Questions 17–21)

17. B The main difference between the mechanisms of action of steroid hormones and peptide hormones is that steroids work intracellularly and peptide hormones work extracellularly. The barrier separating the intracellular environment from the extracellular environment is the lipid bilayer known as the plasma membrane. Peptide hormones must exert their effects extracelluarly, because they are not lipid-soluble and therefore can't easily cross the plasma membrane and enter a cell. Instead, peptide hormones bind to extracellular receptors on the plasma membranes of their target cells; and this binding then triggers a series of enzymatic reactions within the cells. Steroid hormones, on the other hand, are small hormones related to and synthesized from cholesterol, which is a component of eukaryotic plasma membranes. Thus, steroids are lipid-soluble and can easily traverse the plasma membrane, so they can and do act inside their target cells. Once inside a cell, a steroid hormone binds to a cytoplasmic receptor protein. The receptor-steroid complex enters the nucleus, where it induces protein synthesis by binding to the DNA and derepressing specific genes. Thus, choice (B) is the correct answer. Choice (A) doesn't hang together logically, and even if it did, it would not explain the differences between he modes of action of the two types of hormones. It's true that 17-β-estradiol is hydrophobic; however, this property doesn't make it stable in the cytoplasm, which is aqueous. Hydrophobic proteins, in fact, tend to be unstable in the cytoplasm, because their tertiary structure tends to be disturbed by a polar environment; however, a small steroid hormone like 17-β-estradiol has no tertiary structure to lose, so its stability won't be affected one way or the other by the cytoplasmic surroundings. Moreover, if it were true that this molecule was particularly stable in the cytoplasm, this would be a property that it shared with peptide hormones, so it wouldn't explain the differences in modes of action. Choice (C) is wrong because it's simply untrue. There is no need for estradiol to bind to extracellular plasma membrane receptors, because it's lipid-soluble and therefore acts intracellularly. Finally, choice (D) can be eliminated for a number of reasons. First, the passage doesn't give you any information about the size of insulin versus the size of 17-β-estradiol; and although it's true (and you may have known) that proteins are generally larger than steroids, it's not the size of the hormones that dictates their mode of action, but their solubility in the plasma membrane. In fact, as proteins go, insulin is fairly small, and enzymes much larger than insulin normally act in the nucleus, interacting with DNA during DNA replication and protein synthesis; so its size certainly wouldn't preclude insulin from interacting with DNA.

18. C To answer this question, you first have to figure out what kind of influence increasing estradiol levels have on FSH secretion between Assay B and Assay C, which corresponds to the time interval between days 5 and 10. During this interval, the estradiol plasma concentration rose from 49 to 240.2, while the plasma concentration of FSH decreased from 12 to 10. The question stem implies that this FSH decrease is a direct result of the estradiol increase; thus, it appears that this effect is one of negative biofeedback, choice (C). In biological systems there are many modes of regulation, one of which is negative biofeedback. Often, the concentration of a product or intermediate in a metabolic pathway inhibits the pathway that led to its formation. And although this is not exactly the case here, there is a negative relationship between estradiol and FSH; when the concentration of estradiol rises sharply prior to ovulation, it makes sense that FSH secretions should be inhibited, because FSH stimulates the maturation of an ovarian follicle, and it would be metabolically wasteful for a second follicle to mature before the first one is even mature, let alone ovulated. So, choice (C) is the correct answer. Choice (A), positive biofeedback, is another biological regulatory mechanism, in which the increased secretion of one product stimulates the increased secretion of a second product, which is not the effect that estradiol has on FSH production over the time period in question. Choice (B), competitive inhibition, might have sounded like a good choice, since FSH secretion does indeed appear to be inhibited by estradiol, but this inhibition is in no way competitive. Competitive inhibition is a regulatory mechanism in which molecules that are structurally similar compete with one another for substrate binding sites. The passage gives you no evidence at all about the mechanism whereby 17-β-estradiol affects FSH secretion, and certainly gives you no reason to believe that competitive inhibition is occurring; thus, choice (B) is wrong. Finally, choice (D), allosteric activation, is another regulatory mechanism. Allosteric effects occur in molecules with multiple active sites: the binding of substrate at one active site affects the properties of the others. Thus, allosteric activation is when substrate binding increases the reactivity of other active sites. Obviously, this is not occurring here, so choice (D) is wrong.

19. A Ovulation is followed by the conversion of the ruptured follicle to the corpus luteum, which maintains the uterine endometrium through its secretions of estrogens and progesterone. Therefore, it can be said that ovulation is indirectly responsible for the increases in estrogen concentration and progesterone concentration that follow it. So, if breast-feeding stimulates secretion of the hormone prolactin, one of the results being the absence of menstrual periods, then this means that prolactin must inhibit ovulation. Thus, choices (C) and (D) can be eliminated. Since prolactin inhibits ovulation, then it must also inhibit progesterone and estrogen secretion; and you might also have known that estrogen stimulates LH and therefore helps bring on ovulation, which means that ovulation will tend to be inhibited by decreased, not increased, estrogen levels; therefore, choice (B) is wrong and choice (A) is correct.

20. A Answering this question requires a little bit of outside knowledge of the menstrual cycle and its hormonal regulation. Looking at Table 1, you should have either known outright that ovulation occurs midway through the cycle, around Day 14; or, if you recalled that ovulation is preceded by a surge in estrogen and LH secretion, you could have reasoned this out based on the table. The peak in estrogen concentration at Day 10 triggers the rise in LH concentration between Days 10 and 15, and it's this surge of LH secretion that triggers ovulation. Following ovulation, the ruptured ovarian follicle forms a yellowish mass of cells called the corpus luteum; as mentioned in the passage, this begins to secrete progesterone and estrogens, evoking the increase in progesterone and 17-β-estradiol secretion that takes place between Assay E and Assay F.

So, choice (A) is the correct answer. Even if you didn't know this, you might have been able to eliminate the other three choices. Choice (B) can be eliminated based on logic: it claims that FSH-induced inhibition of the enzyme that catalyzes the formation of progesterone from its precursor, pregnenalone, is responsible for the rise in progesterone and 17-β-estradiol secretion. Besides the fact that there's no evidence for any such mechanism in the passage, if this were in fact true, then you'd expect the progesterone concentration to decrease following the surge in FSH secretion on Day 15. You might also expect a significant decrease in FSH concentration around Day 23, when the progesterone level surges. Since neither of these is observed, choice (B) can be ruled out. On the other hand, if choice (C) were correct, it would explain the data in Table 1; however, it's untrue that the uterine glands secrete progesterone and 17-β-estradiol in anticipation of implantation. As discussed earlier, it's the corpus luteum that secretes these two hormones. As a matter of fact, although you might not have known this, the only glands that secrete estrogens and progesterone in the uterus are those of the placenta, but this only happens if pregnancy occurs. As for choice (D), although LH does have a biofeedback relationship with estrogen and progesterone, which at times is one of negative biofeedback, and at other times positive biofeedback, you should have immediately ruled out this choice on a technicality: if you look at Table 1, you'll see that LH secretion decreases between Days 15 and 19.

21. D Pregnenalone is converted into progesterone; the question asks you to describe the change that occurs during this conversion. To answer this, you need to look at the structural differences between the two compounds, and also remember that a loss of hydrogen atoms is known as oxidation, while the opposite process, a gain of hydrogen atoms, is known as reduction. We can see that there are only two major changes from one structure to the other: the double bond in the second ring of pregnenalone is shifted into the first ring, and the —OH, or hydroxy group, in pregnenalone becomes a double-bond-O, or carbonyl group, in progesterone. The conversion of a hydroxy group into a carbonyl group is brought about by removal of hydrogen atoms, or oxidation. Therefore, choice (D), oxidation of a hydroxy group, is correct, and choice (C) is incorrect. Choice (A), reduction of a carbon-carbon single bond, is incorrect because the only reduction occurs at the carbon-carbon double bond in the second ring of pregnenalone. Here, hydrogen atoms are added to the double bond, making it a single carbon-carbon bond as the double bond shifts over to the first ring. In fact, a carbon-carbon single bond can't be reduced. Choice (B), oxidation of a double bond, is incorrect because what gets oxidized is a carbon-carbon single bond; and as we said, the double bond got reduced, not oxidized. Oxidation of a double bond would result in the removal of more hydrogen atoms and give a triple bond. Since there's no triple bond in progesterone, choice (B) is clearly incorrect.

Passage IV (Questions 22–28)

22. D The DNA fragment is labeled with radioactive phosphorus-32 to identify the polarity of the DNA fragment—that is, which is the 5' end and which is the 3' end. The radioactive phosphorus is inserted into phosphate which is then incorporated into DNA during replication since phosphate makes up the backbone of the double helix. Therefore, when fragments of DNA are subjected to autoradiography following polyacrylamide gel electrophoresis, the 5' end, which is labeled, can be distinguished from the unlabeled 3' end which the passage says is ignored. So, choice (D) is correct. Even if you didn't see this, you could get the correct answer by a process of elimination, based on the descriptions of the procedure. According to the passage, the DNA is first radiolabeled, then divided into four parts, and then each part is subjected to a different chemical treatment. Thus, choice (A), which suggests that the radiolabeling causes the cleavage, is incorrect. And so is choice (B), which suggests that the radiolabeling itself functions to divide the sample into the four equal parts. Choice (C) is incorrect since radiolabeling merely replaces a nonradioactive phosphorus atom with a radioactive phosphorus atom; this does not change the reactivity of the molecule and therefore does not have any effect on the duration of the cleavage reaction. The duration presumably depends only on the amount of time the DNA sample is exposed to the reagent that causes cleavage.

23. C If a DNA-RNA hybrid were to be formed with a DNA fragment having the sequence AGACCTATG in the 5' to 3' direction, then the sequence of the RNA fragment would have to be complementary to the sequence of the DNA. There are two things you have to keep in mind: first, in RNA, uracil takes the place of thymine; that is, adenine binds with uracil. Second, in double-stranded nucleic acid, such as in a DNA helix or a DNA-RNA hybrid, the two strands run antiparallel; that is, the 5' end of the DNA binds with the 3' end of the RNA. Therefore, the sequence of the RNA fragment complementary to the DNA fragment would be CAUAGGUCU in the 5' to 3' direction, which means that choice (C) is the right answer. You should have ruled out choices A and B immediately, since they both contain thymine, and RNA does not have thymine. Choice (D) is wrong because its polarity is wrong: its sequence is backwards.

24. C If a band appeared at the same position in both the G lane and the A + G lane, then the base would be identified as guanine. One of the chemical reactions cleaves before either purine—that is, before adenine or before guanine. Therefore, for every adenine and every guanine a band will appear in the A + G lane. However, this band alone is not enough to distinguish between the two purines. One of the other chemical reactions cleaves the fragment only before guanines, and so for every guanine a band will also appear in the G lane. Logically, therefore, if there is a band at the same position in both the G lane and the A + G lane, then the base can be conclusively identified as a guanine. Thus, choice (C) is correct and choice (D) is wrong. Choice (B) is also wrong, because by the same reasoning, a band that only appears in the A + G lane would be identified as an adenine. Finally, choice (A), the G lane only, is incorrect, since for every guanine, a band will appear in the G lane and in the A + G lane. In fact, you would never see a band only in the G lane, unless there was some sort of error in your procedure.

25. B The DNA fragment in Figure 1 has the sequence 5' GACACTCAAG 3'. There are some clues to help you determine the proper sequence. First of all, the passage tells you that the band appearing in all four lanes at the top of the gel is unreacted DNA fragment. Second, the polarity of the gel is conveniently labeled for you: the 5' end is at the bottom of the lane, and the 3' end is at the top. So to read the sequence in the 5' to 3' direction, which you want to do since all the answer choices have this polarity, you have to work from the bottom up. Bands appearing in the G lane represent cleavage before a guanine; bands appearing in the C lane represent cleavage before a cytosine; bands appearing in the A + G lane represent cleavage before either a guanine or an adenine; and bands appearing in the T + C lane represent cleavage before either a thymine or a cytosine. If a band in the T + C lane has a corresponding band—that is, in the same position—in the C lane, then the cleavage was before a cytosine; if not, then the cleavage was before a thymine. If a band in the A + G lane has a corresponding band in the G lane, then the cleavage was before a guanine; if not, then the cleavage was before an adenine. Thus, reading from the bottom of the lanes to the top, the sequence of the DNA fragment in the 5' to 3' direction is GACACTCAAG, and choice (B) is correct.

26. A Two of the four chemical treatments used in the Maxam-Gilbert method cleave a DNA fragment before TWO different bases—in one case, before either adenine or guanine, and in the other, before either thymine or cytosine. Therefore choice (A), that each of the treatments cleaves the DNA before only one nitrogenous base, is false and thus is not a basic premise underlying the method. So, choice (A) is the right answer. Choices B, C, and D are all incorrect choices since they ARE basic premises underlying the procedure. Choice (B), that DNA contains four nitrogenous bases—adenine, guanine, thymine and cytosine—is a very basic fact about DNA that is a basis for this procedure, since these are the known points at which the fragment of DNA is cleaved by the four chemical treatments. Choice (C), that the lengths of small fragments of DNA must be comparable by standard biochemical means, is also true: size comparison of the fragments in each of the four samples, which is effected by the gel electrophoresis step, is also necessary for determining the base sequence. So choice (C) is also a basic premise. Finally, choice (D) states that a single chemical procedure can produce labeled fragments of varying lengths. The passage tells us that the single chemical procedure for cleaving guanine produces a series of fragments of different lengths, each ending before a different guanine in the original molecule. This allows one to determine all of the positions occupied by a particular base along the DNA molecule. So this is also a basic premise of the Maxam-Gilbert method and choice (D) is also wrong.

27. B According to the passage, if the A + G chemical treatment were carried to completion, the DNA fragment would be cleaved at every adenine and guanine present. This means that on average the original fragment would be cleaved many times, not just once. These DNA fragments would therefore have small molecular weights more similar to one another than the weights of the two fragments produced by a single cleavage of the DNA. Since gel electrophoresis separates the DNA fragments into distinct bands on the basis of molecular weight, a series of fragments with similar molecular weights would produce a "fuzzy" band, as opposed to the more distinct band produced by the single cleavage DNA fragments. Therefore, choice (B) is correct and choice (A) is wrong. If you weren't quite clear on the concept behind this question, you might have looked at choices C and D first, since they can be easily eliminated. Choice (C) is wrong on a technicality, since there is no "A" lane present on the gel;

while choice (D) is incorrect because unlabeled fragments do not show up on the developed autoradiogram of the gel electrophoresis, since they contain no radioactive phosphorus to expose the film.

28. C If the DNA fragment given in the question stem were subjected to the Maxam-Gilbert procedure, it would produce all five of the fragments listed. However, you're interested only in those fragments that would produce bands in the C lane or the T + C lane of the gel electrophoresis. Between them, the two chemical treatments that produce bands in the C lane and the T + C lane cleave before cytosine residues and thymine residues. Cleavage before cytosine residues would produce only one fragment, (32P)A; while cleavage before thymine residues would produce two fragments, (32P)AC and (32P)ACTA. This corresponds to fragments I, II, and IV, and so choice (C) is correct.

Discrete Questions

29. B This is another knowledge-based question, requiring you to distinguish between biological processes that require ATP and those that don't. Exocytosis of synaptic vesicles containing neurotransmitters at a nerve terminal is an ATP-dependent process that is triggered by the transmission of an action potential along the length of a neuron; so choice (A) is wrong. Urea, a by-product of amino acid metabolism, is a small uncharged molecule and therefore can cross the cell membrane by simple diffusion, which is a passive process. Therefore, choice (B) is independent of ATP and is thus the correct answer. Calcium *enters* muscle cells by flowing down its electrochemical gradient—ATP is not required. (In skeletal muscle it is released from the SR and in smooth muscle it flows in through a voltage-gated Ca^{++} channel.) The movement of calcium *OUT* of a muscle cell goes against its concentration gradient and therefore requires ATP. Choice (D), the export of sodium ions from a neuron, occurs in conjunction with the import of potassium ions; this is known as the sodium-potassium-ATPase pump, an ATP-dependent process necessary for the maintenance of a potential across the neuron membrane. Therefore, choice (D) is wrong.

30. C This question is concerned with the kinetics of an enzyme catalyzed reaction. Enzymes speed up the rates of reactions that would eventually occur on their own in due time, by decreasing the activation

energy without being consumed by the reaction. The activation energy is the amount of energy that the reactants must absorb from their surroundings to reach the transition state. In an exothermic reaction, such as the one depicted in the figure, the free energy of the reactants is greater than the free energy of the product. The initial free energy of the reactants and the final free energy of the products are independent of enzyme catalysis. Therefore, if the enzyme in question were blocked by an inhibitor, the free energy of the reactants and the products would remain the same—which means that choices A and B are wrong—but the activation energy would increase. This means that choice (D) is clearly wrong and choice (C) is the correct answer. You might have realized that the graph is superfluous in terms of answering this questions.

Passage V (Questions 31–35)

31. B Identify which of the choices is not found in ANY bacteria. Bacteria are prokaryotes, and that the main characteristics of prokaryotes are that they have circular DNA, they don't have ANY membrane-bound organelles, they have ribosomes (which are structurally different from eukaryotic ribosomes), and they have cell walls composed of complex macromolecules of amino acids and amino sugars. Thus, choices (A), (C), and (D) are structures found in all bacteria, and thus can be eliminated. Choice (B), mitochondria, are the membrane-bound organelles that supply eukaryotic cells with ATP, and so choice (B) is the correct answer.

32. C Choice (D) is the odd man out, so it's worth checking quickly to see if it's correct or not. In fact, chemoautotrophs are organisms that derive their energy from the oxidation of inorganic chemical compounds rather than organic compounds, and require carbon dioxide for growth; there's no reason to believe that this describes Strain 2; in fact, you have reason to believe that it doesn't, since Strain 2 can digest some starches, which are organic molecules. So choice (D) is incorrect. As for the other three: according to the results of Experiment 2, Strain 2 exhibited colony growth both on the surface of the agar, which is exposed to oxygen, and within the agar itself, which according to the passage, is oxygen-poor. To figure out whether this fits choice (A), (B), or (C), you have to know or figure out what these terms mean. An obligate anaerobe is an organism that obtains its energy via anaerobic processes such as fermentation. Since oxygen is a highly reactive compound, and since anaerobes don't consume oxygen in metabolism, oxygen tends to be toxic to anaerobic organisms. Thus, an obligate anaerobe would not be expected to grow on the surface of an agar plate. Since Strain 2 did exhibit growth under these conditions, choice (A) is incorrect. Obligate aerobes, choice (B), are organisms that require oxygen for metabolism; this implies that such an organism would not exhibit growth within an oxygen-poor environment such as the inside of the agar layer in Experiment 2. Thus, choice (B) is also incorrect. This leaves choice (C), facultative anaerobes. A facultative anaerobe is an organism that normally derives its energy aerobically, but also has metabolic pathways that allow it to exist under anaerobic conditions, such as within the layer of agar. Since this definition corresponds to the results of Experiment 2, Strain 2 bacteria would most likely be classified as facultative anaerobes, and so choice (C) is correct.

33. B The results of Experiment 1 for Strain 1 indicate that Strain 1 is capable of growing on all three of the starch-agar plates used in the protocol, though iodine staining revealed that, at least for the first 48 hours of growth, Strain 1 does not digest Starch A, B, or C. In other words, although starch digestion is absent during the first 48 hours, colony growth occurs. This implies that Strain 1 bacteria must not use starch for its first 48 hours of growth, and so choice (B) must be correct. Choice (A), which says that Strain 1 bacteria do not possess the enzymes necessary for starch digestion, is tempting, since you know that starch digestion doesn't occur in the course of the experiment. However, metabolic pathways are not necessarily active at all times, and it's quite common for a bacterium to use one nutrient source preferentially over another and only to switch to a second source, and the pathways required to utilize it, after the first, more attractive nutrient has been exhausted. So, although it might be the case that the strain doesn't possess a starch digestion pathway, as a researcher, you would not be justified in drawing this conclusion after only 48 hours of incubation time; it could just be that the starch pathway is there but just isn't activated within 48 hours, because there are enough other nutrients present to keep the cells alive for that time. In order to conclusively demonstrate that Strain 1 lacks the enzymatic machinery to digest starch, you would want to repeat Experiment 1 with a longer incubation time. Choice (C) says that Strain 1 bacteria grows best in an oxygenated environment. Although

this claim was actually substantiated by the results of Experiment 2, this conclusion can't be drawn from Experiment 1, because plates A, B, and C are only examined for growth on the agar surface, not within the agar layer. Finally, choice (D) is incorrect because not only is it false that Strain 1 bacteria can't grow on starch-agar medium C—because they do—but even if it were true, this alone would not be enough to account for the discrepancy between starch digestion and colony growth, which is the issue that this question is actually addressing.

34. C This question investigates the mystery of how Strain 1 bacteria can grow on starch-agar media for 48 hours without actually digesting any of the starch. You're told that when the researcher repeated Experiment 1, the results were identical to those described in the passage, except that Strain 1 bacteria now exhibited starch digestion on all three of the starch-agar plates. So, you're asked to decide what conclusions might be drawn from this new data. Statement I says that Strain 1 bacteria require longer incubation times to digest starch. Based on the data, this conclusion does follow: with an incubation period of 120 hours, not only did Strain 1 grow on all three media—which we'd already seen from the original experiment 1—but it also digested all three types of starch. Statement I is correct, so choice (B), III only, can be eliminated. Statement II says Strain 2 needs oxygen for its early stages of development. This is clearly incorrect, because Experiment 1 did not address the oxygen metabolism of the bacterial strains—this was only tested in Experiment II. According to Table 2, Strain 2 grows in the oxygen-poor environment below the agar surface, so it doesn't need oxygen to grow. Thus, Statement II is false, and so choice (D) can also be ruled out. Finally, Statement III says that Strain 2 cannot digest the starch in starch-agar medium C. Based on this, it is fairly safe to assume that if Strain 2 has not started to grow after this length of time, then it won't grow at all. Thus, Statement III is a valid conclusion, which means that choice (A) is wrong and choice (C) is the right answer.

35. A In this version of the experiments, Strain 2 exhibited starch digestion on all three of the starch-agar media, whereas in the first run of the experiments, Strain 2 digested only two out of the three different starches—it didn't digest Starch C. This means that between the first and second trials, something happened to some of the bacteria in Strain 2 that now allowed them to digest Starch C. Mutation, choice (A), is a change in DNA sequence; though most mutations are deleterious to an organism, mutations sometimes have beneficial results. An example of this would be if a mutation made Strain 2 able to digest Starch C in the second trial of Experiment 1, since this ability would increase the bacteria's survival ability. A mutation that resulted in the synthesis of an enzyme capable of digesting Starch C could plausibly explain the observed results. Thus, choice (A) is the correct answer. Transduction, choice (B), is the transfer of bacterial DNA between two bacteria via a bacteriophage, which is a type of virus that only infects bacteria. Though transformation could account for Strain 2's newfound ability to digest Starch C, since the introduction of new DNA might provide the bacteria with some enzymes of metabolic pathways that it previously lacked, this is not the most likely explanation in this case, because there's no evidence of bacteriophage infection in the bacterial strains. Furthermore, there's no bacterial strain around that can digest Starch C from which the capacity could be transformed. So, choice (B) is wrong. Choice (C), nondisjunction, is the failure of paired chromosomes or chromatids to properly separate during mitosis or meiosis, resulting in daughter cells that either lack a chromosome, or have triplicate copies of one. Since bacteria don't have separate chromosomes but just one piece of circular DNA, nondisjunction can't possibly occur during bacterial DNA replication. And, even if the chromosome failed to replicate correctly, this wouldn't produce any new DNA, so it couldn't possibly give the bacteria a new metabolic pathway, and choice (C) is wrong. Finally, choice (D) is wrong because meiosis is the eukaryotic process by which gametes are formed. Prokaryotic organisms do not undergo meiosis at any stage of their existence; they replicate via binary fission, which is basically mitosis.

Discrete Questions

36. B You're required to know what structures and systems each of the three embryonic germ layers gives rise to. Spina bifida is a bone abnormality; since bone arises from the mesodermal embryonic germ layer, a lesion to the mesoderm that simulates this disease would most likely also affect the development of other structures based on different types

of connective tissue, such as blood, blood vessels and the muscles and diffuse connective tissue of various organs. Thus, this lesion would affect the development of both muscles and blood vessels, Roman Numerals I and II, and so choice (A) is wrong and choice (B) is correct. Skin and hair, along with the nervous system, originate from ectoderm, while intestinal develops from endoderm. Therefore, Roman Numerals III and IV are wrong, and thus choices (C) and (D) are also wrong.

37. D Oxygen and hydrogen ions are in an allosteric relationship with respect to hemoglobin. An increase in the concentration of hydrogen ions decreases hemoglobin's affinity for oxygen; that is, the binding of hydrogen ions to a molecule of oxyhemoglobin enhances the release of oxygen. This interaction is known as the Bohr effect. According to the question, a patient suffering from acidosis has a decreased blood pH, which means that the concentration of hydrogen ions in the blood is higher than normal. And, as just discussed, a high concentration of hydrogen ions means that hemoglobin will release oxygen more readily than it does under normal conditions. With reference to the hemoglobin dissociation curve, this means that at a given partial pressure of oxygen in the blood, the percent saturation of hemoglobin with oxygen in a patient with acidosis will be lower than it would be at physiological pH. In terms of the graph, the hemoglobin dissociation curve corresponding to an acidotic patient would be shifted to the right of the curve for normal pH, and so curve D, choice (D), is correct.

38. A Pepsin is an enzyme secreted by the chief cells of the stomach's gastric glands; it digests proteins by hydrolyzing specific peptide bonds. In addition to pepsin, the gastric glands secrete hydrochloric acid, which give the stomach a pH of about 2. Thus, since the stomach is very acidic, and since pepsin functions in that environment, it follows that its pH optimum will be around 2; in other words, a graph of its activity as a function of pH would peak at a pH of 2, and slope downward as pH increases. Thus, choice (A) is correct. The reason pepsin works best in such an acidic environment is because the low pH of gastric juice denatures the proteins found in food, thereby exposing their peptide bonds to the enzyme's actions.

MCAT Biological Sciences Test

Time—100 minutes

Questions 1–74

DIRECTIONS: Most of the questions in the following Biological Sciences test are organized into groups, with a descriptive passage preceding each group of questions. Study the passage, then select the single best answer to each question in the group. Some of the questions are not based on a descriptive passage; you must also select the best answer to these questions. If you are unsure of the best answer, eliminate the choices that you know are incorrect, then select an answer from the choices that remain. Indicate your selection by blackening the corresponding circle on your answer sheet. A periodic table is provided below for your use with the questions.

Passage I (Questions 1–7)

Hemoglobin (Hb) and myoglobin (Mb) are the O_2-carrying proteins in vertebrates. Hb, which is contained within red blood cells, serves as the O_2 carrier in blood and also plays a vital role in the transport of CO_2 and H^+. Vertebrate Hb consists of four polypeptides (subunits), each with a heme group. The four chains are held together by noncovalent attractions. The affinity of Hb for O_2 varies between species and within species depending on such factors as blood pH, stage of development, and body size. For example, small mammals give up O_2 more readily than large mammals because small mammals have a higher metabolic rate and require more O_2 per gram of tissue.

PERIODIC TABLE OF THE ELEMENTS

1 H 1.0																	2 He 4.0
3 Li 6.9	4 Be 9.0											5 B 10.8	6 C 12.0	7 N 14.0	8 O 16.0	9 F 19.0	10 Ne 20.2
11 Na 23.0	12 Mg 24.3											13 Al 27.0	14 Si 28.1	15 P 31.0	16 S 32.1	17 Cl 35.5	18 Ar 39.9
19 K 39.1	20 Ca 40.1	21 Sc 45.0	22 Ti 47.9	23 V 50.9	24 Cr 52.0	25 Mn 54.9	26 Fe 55.8	27 Co 58.9	28 Ni 58.7	29 Cu 63.5	30 Zn 65.4	31 Ga 69.7	32 Ge 72.6	33 As 74.9	34 Se 79.0	35 Br 79.9	36 Kr 83.8
37 Rb 85.5	38 Sr 87.6	39 Y 88.9	40 Zr 91.2	41 Nb 92.9	42 Mo 95.9	43 Tc (98)	44 Ru 101.1	45 Rh 102.9	46 Pd 106.4	47 Ag 107.9	48 Cd 112.4	49 In 114.8	50 Sn 118.7	51 Sb 121.8	52 Te 127.6	53 I 126.9	54 Xe 131.3
55 Cs 132.9	56 Ba 137.3	57 La* 138.9	72 Hf 178.5	73 Ta 180.9	74 W 183.9	75 Re 186.2	76 Os 190.2	77 Ir 192.2	78 Pt 195.1	79 Au 197.0	80 Hg 200.6	81 Tl 204.4	82 Pb 207.2	83 Bi 209.0	84 Po (209)	85 At (210)	86 Rn (222)
87 Fr (223)	88 Ra 226.0	89 Ac† 227.0	104 Unq (261)	105 Unp (262)	106 Unh (263)	107 Uns (262)	108 Uno (265)	109 Une (267)									

	58 Ce 140.1	59 Pr 140.9	60 Nd 144.2	61 Pm (145)	62 Sm 150.4	63 Eu 152.0	64 Gd 157.3	65 Tb 158.9	66 Dy 162.5	67 Ho 164.9	68 Er 167.3	69 Tm 168.9	70 Yb 173.0	71 Lu 175.0
†	90 Th 232.0	91 Pa (231)	92 U 238.0	93 Np (237)	94 Pu (244)	95 Am (243)	96 Cm (247)	97 Bk (247)	98 Cf (251)	99 Es (252)	100 Fm (257)	101 Md (258)	102 No (259)	103 Lr (260)

The binding of O_2 to Hb is also dependent on the co-operativeness of the Hb subunits. That is, binding at one heme facilitates the binding of O_2 at the other hemes within the Hb molecule by altering the conformation of the entire molecule. This conformational change makes subsequent binding of O_2 more energetically favorable. Conversely, the unloading of O_2 at one heme facilitates the unloading of O_2 at the others by a similar mechanism.

Figure 1 depicts the O_2-dissociation curves of Hb (Curves A, B, and C) and Myoglobin (Curve D), where saturation, Y, is the fractional occupancy of the O_2-binding sites.

The fraction of O_2 that is transferred from Hb as the blood passes through the tissue capillaries is called the *utilization coefficient*. A normal value is approximately 0.25.

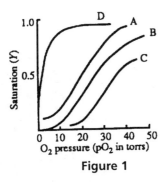

Figure 1

Myoglobin facilitates O_2 transport in muscle and serves as a reserve store of O_2. Mb is a single polypeptide chain containing a heme group, with a molecular weight of 18 kd. As can be seen in Figure 1, Mb (Curve D) has a greater affinity for O_2 than Hb.

1. The llama is a warm-blooded mammal that lives in regions of unusually high altitudes, and has evolved a type of Hb that adapts it to such an existence. If Curve B represents the O_2-dissociation curve for horse Hb, which curve would most closely resemble the curve for llama Hb?

   A. Curve A
   B. Curve B
   C. Curve C
   D. Curve D

2. If Curve B represents the O_2-dissociation curve for elephant Hb, which curve most closely resembles the curve for mouse Hb?

   A. Curve A
   B. Curve B
   C. Curve C
   D. Curve D

3. If Curve B represents the O_2-dissociation curve for human adult Hb, which of the following best explains why Curve A most closely resembles the curve for fetal Hb?

   A. Fetal tissue has a higher metabolic rate than adult tissue.
   B. Fetal tissue has a lower metabolic range than adult tissue.
   C. Fetal Hb has a higher affinity for O_2 than adult Hb.
   D. Fetal Hb has a lower affinity for O_2 than adult Hb.

4. The sigmoidal shape of the O_2-dissociation curve of Hb is due to:

   A. the effects of oxidation and reduction on the heme groups within the Hb molecule.
   B. the concentration of carbon dioxide in the blood.
   C. the fact that Hb has a lower affinity for O_2 than Mb.
   D. the cooperativity in binding among the sub-units of the Hb molecule.

5. A sample of human adult Hb is placed in an $8M$ urea solution, resulting in the disruption of non-covalent interactions. After this procedure, the α chains of Hb are isolated. Which of the four curves most closely resembles the O_2-dissociation curve for the isolated α chains? [Note: Assume that Curve B represents the O_2-dissociation curve for human adult Hb *in vivo*.]

   A. Curve A
   B. Curve B
   C. Curve C
   D. Curve D

6. The utilization coefficient is continually being adjusted in response to physiological changes. Which of the following values most likely represents the utilization coefficient for human adult Hb during strenuous exercise?

   A. 0.0
   B. 0.125
   C. 0.25
   D. 0.75

7. In sperm whales, the Mb content of muscle is about 0.004 mole/kg of muscle. If a sperm whale has 1,000 kg of muscle, approximately how much O_2 is bound to Mb, assuming that the Mb is saturated with O_2?

   A. 4 moles
   B. 8 moles
   C. 12 moles
   D. 16 moles

Passage II (Questions 8–14)

Just as the ingestion of nutrients is mandatory for human life, so is the excretion of metabolic waste products. One of these nutrients, protein, is used for building muscle, nucleic acids, and countless compounds integral to homeostasis. However, the catabolism of the amino acids generated from protein digestion produces ammonia, which if not further degraded, can become toxic. Similarly, if the same salts that provide energy and chemical balance to cells are in excess, fluid retention will occur, damaging the circulatory, cardiac, and pulmonary systems.

One of the most important homeostatic organs is the kidney, which closely regulates the excretion and reabsorption of many essential ions and molecules. One mechanism of renal function involves the secretion of antidiuretic hormone (ADH).

Diabetes insipidus (DI), is the condition that occurs when ADH is ineffective. As a result, the kidneys are unable to concentrate urine, leading to excessive water loss. There are two types of DI: central and nephrogenic. *Central DI* occurs when there is a deficiency in the quantity or quality of ADH produced. *Nephrogenic DI* occurs when the kidney tubules are unresponsive to ADH. To differentiate between these two conditions, a patient's urine osmolarity is measured both prior to therapy and after a 24-hour restriction on fluid intake. Exogenous ADH is then administered and urine osmolarity is measured again. The table below gives the results of testing on four patients. Assume that a urine osmolarity of 285 mOsm/L of H_2O is normal.

Table 1

Urine Osmolarity (mOsm/L of H_2O)

Patient	Before therapy	After fluid restriction	After ADH
A	285	765	765
B	180	765	765
C	180	180	400
D	180	180	180

8. An elevated and potentially toxic level of ammonia in the blood (*hyperammonemia*) would most likely result from a defect in an enzyme involved in:

    A. glycolysis.
    B. fatty acid catabolism.
    C. the urea cycle.
    D. nucleic acid degradation.

9. According to the passage, the catabolism of amino acids produces ammonia. Therefore, after a protein-rich meal, would you expect a buildup of ammonia in the lumen of the small intestine?

    A. Yes, because the ammonia will not be able to diffuse into the intestinal epithelium.
    B. Yes, because the rate at which digestive enzymes degrade ammonia is slower than the rate at which ammonia is produced.
    C. No, because the ammonia will diffuse into the intestinal epithelium and will be excreted by the kidneys.
    D. No, because the ammonia is produced inside individual cells, not within the lumen of the small intestine.

10. Which of the following substances would NOT be found in appreciable quantity in the urine of a healthy individual?

    A. Albumin
    B. Sodium
    C. Urea
    D. Potassium

11. Which of the following would you most likely expect to find in a patient with diabetes insipidus?

    A. Decreased plasma osmolarity
    B. Increased urine osmolarity
    C. Increased urine glucose
    D. Increased urine output

12. Based on the data in Table 1, which of the four patients most likely has central diabetes insipidus?

    A. Patient A
    B. Patient B
    C. Patient C
    D. Patient D

13. Based on the data in Table 1, which of the four patients most likely has nephrogenic diabetes insipidus?

    A. Patient A
    B. Patient B
    C. Patient C
    D. Patient D

14. What is the most likely cause of Patient B's dilute urine before therapy?

    A. Excessive water intake
    B. Dehydration
    C. Nephrogenic DI
    D. Central DI

Questions 15 through 17 are NOT based on a descriptive passage.

15. An increase in heart rate, blood pressure, and blood glucose concentration are all associated with stimulation of the:

    A. parasympathetic nervous system.
    B. sympathetic nervous system.
    C. somatic nervous system.
    D. digestive system.

16. Which of the following structures plays a role in both the male excretory and reproductive systems, but in the female excretory system only?

    A. Epididymis
    B. Prostate
    C. Urethra
    D. Ureter

17. Which of the following cell types does NOT contain the diploid number of chromosomes?

    A. Spermatogonium
    B. Spermatid
    C. Zygote
    D. Primary oocyte

Passage III (Questions 18–22)

Hemophilia is a genetically inherited disease that causes the synthesis of an abnormal clotting factor. As a result, hemophiliacs bleed excessively from the slightest injury. The figure below is a partial pedigree for the hemophilia trait in Queen Victoria's descendants. The pedigree indicates no history of hemophilia for either parent prior to the F_1 generation.

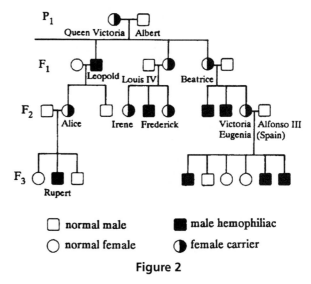

Figure 2

18. According to Figure 2, which of the following assumptions about the P_1 generation must be true?

    A. Albert did not have the gene for hemophilia.
    B. Queen Victoria had two X chromosomes, each with the gene for hemophilia.
    C. Neither Albert nor Queen Victoria had the gene for hemophilia.
    D. Albert was a carrier of the hemophilia gene.

19. Which of the following best explains why Louis IV was NOT a hemophiliac?

    A. His son Frederick was a hemophiliac.
    B. He did not inherit the gene for hemophilia from his mother.
    C. His father-in-law, Albert, was not a hemophiliac.
    D. Only females can be carriers of the gene for hemophilia.

20. If Beatrice had married a hemophiliac and had a son, what is the probability that the son would have been a hemophiliac?

    A. 0%
    B. 25%
    C. 50%
    D. 100%

21. Theoretically, what percentage of Victoria Eugenia's sons should have been hemophiliacs?

    A. 25%
    B. 33%
    C. 50%
    D. 75%

22. Based on the pedigree, what is the most reasonable explanation for Rupert's hemophilia?

    A. A mutation occurred on the Y chromosome that he inherited from his father.
    B. His mother was a hemophiliac and transmitted the gene to him.
    C. His father was a carrier of the gene for hemophilia.
    D. His maternal grandfather was a hemophiliac.

Passage IV (Questions 23–29)

Aerobic respiration is the major process used by oxygen-requiring organisms to generate energy. During respiration, glucose is metabolized to generate chemical energy in the form of ATP:

$$C_6H_{12}O_6 + 6\ O_2 \rightarrow 6\ CO_2 + H_2O + 36\ ATP$$

The biochemical machinery necessary for cellular respiration is found in the mitochondria, small organelles scattered throughout the cytoplasm of most eukaryotic cells. The number of mitochondria per cell varies by tissue type and cell function.

Mitochondria are unusual in that they have their own genetic systems that are entirely separate from the cell's genetic material. However, mitochondrial replication is still dependent upon the cell's nuclear DNA to encode essential proteins required for replication. Despite this fact, mitochondria seem to replicate randomly, out of phase with both the cell cycle and other mitochondria.

The nature of the mitochondrial genome and protein-synthesizing machinery has led many researchers to postulate that mitochondria may have arisen as the result of the ingestion of a bacterium by a primitive cell millions of years ago. It is postulated that the two may have entered into a symbiotic relationship and eventually become dependent on one another; the cell sustained the bacterium, while the bacterium provided energy for the cell. Gradually, the two evolved into the present-day eukaryotic cell, with the mitochondrion retaining some of its own DNA. This is known as the *endosymbiotic hypothesis*. Because mitochondrial DNA is inherited in a non-Mendelian fashion (mitochondria are inherited from the maternal parent, who supplies most cytoplasm to the fertilized eggs), it has been used to look at evolutionary relationships among different organisms.

23. In which of the following phases of the cell cycle could mitochondrial DNA replicate?

    I. G1
    II. S
    III. G$_2$
    IV. M

    A. IV only
    B. I and III only
    C. II and IV only
    D. I, II, III, and IV

24. Scientists have demonstrated that human mitochondrial DNA mutates at a fairly slow rate. Because mitochondria play such an important role in the cell, these mutations are most likely to be:

    A. point mutations.
    B. frameshift mutations.
    C. lethal mutations.
    D. nondisjunctions.

25. Which of the following mitochondrial genome characteristics differs most from the characteristics of the nuclear genome?

    A. Mitochondrial DNA is a double helix.
    B. Some mitochondrial genes code for tRNA.
    C. Specific mutations to mitochondrial DNA can be lethal to the organism.
    D. Almost every base in mitochondrial DNA codes for a product.

26. What is the net number of ATP molecules synthesized by an obligate anaerobe per molecule of glucose?

    A. 2 ATP
    B. 6 ATP
    C. 8 ATP
    D. 36 ATP

27. A mating type of wild-type strain of the algae *C. reinhardii* is crossed with the opposite mating type of a mutant strain of the algae, which has lost all mitochondrial functions due to deletions in its mitochondrial genome. All of the offspring from this cross also lack mitochondrial functions. Based on information in the passage, this can best be explained by the:

    A. endosymbiotic hypothesis.
    B. non-Mendelian inheritance of mitochondrial DNA.
    C. recombination of mitochondrial DNA during organelle replication.
    D. presence of genetic material in the mitochondria that is distinct from nuclear DNA.

28. Four different human cell cultures—erythrocytes, epidermal cells, skeletal muscle cells, and intestinal cells—were grown in a medium containing radioactive adenine. After 10 days, the mitochondria were isolated via centrifugation, and their level of radioactivity was measured using a liquid scintillation counter. Which of the following cells would be expected to have the greatest number of counts per minute of radioactive decay?

    A. Erythrocytes
    B. Epidermal cells
    C. Skeletal muscle cells
    D. Intestinal cells

29. Which of the following pieces of evidence would NOT support the hypothesis that mitochondria were once independent bacteria that eventually formed a symbiotic relationship with eukaryotic cells?

    A. Mitochondrial DNA is circular and not enclosed by a nuclear membrane.
    B. Mitochondrial ribosomes more closely resemble eukaryotic ribosomes than prokaryotic ribosomes.
    C. Many present-day bacteria live within eukaryotic cells, digesting nutrients that their hosts cannot and sharing the energy thus derived.
    D. Mitochondrial DNA codes for its own ribosomal RNA.

Passage V (Questions 30–37)

Four major blood types exist in the human ABO blood system: types A, B, AB, and O; and there are three alleles that code for them. The A and B alleles are codominant, and the O allele is recessive. Blood types are derived from the presence of specific polysaccharide antigens that lie on the outer surface of the red blood cell membrane. The A allele codes for the production of the A antigen; the B allele codes for the production of the B antigen; the O allele does not code for any antigen.

While there are many other antigens found on red blood cell membranes, the second most important antigen is the Rh antigen. Rh is an autosomally dominant trait coded for by 2 alleles. If this antigen is present, an individual is Rh$^+$; if it is absent, an individual is Rh$^-$. For example, a person with AB blood with the Rh antigen is said to be AB$^+$.

These antigens become most important when an individual comes into contact with foreign blood. Because of the presence of naturally occurring substances that closely mimic the A and B antigens, individuals who do not have these antigens on their red blood cells will form antibodies against them. This is inconsequential until situations such as blood transfusion, organ transplant, or pregnancy occur.

Erythroblastosis fetalis is a condition in which the red blood cells of an Rh$^+$ fetus are attacked by antibodies produced by its Rh$^-$ mother. Unlike ABO incompatibility, in which there are naturally occurring antibodies to foreign antigens, the Rh system requires prior sensitization to the Rh antigen before antibodies are produced. This sensitization usually occurs during the delivery of an Rh$^+$ baby. So while the first baby will not be harmed, any further Rh$^+$ fetuses are at risk.

The *Coombs tests* provide a method for determining whether a mother has mounted an immune response against her baby's blood. The tests are based on whether or not agglutination occurs when Coombs reagent is added to a sample. Coombs reagent contains antibodies against the anti-Rh antibodies produced by the mother. The *indirect Coombs test* takes the mother's serum, which contains her antibodies but no red blood cells, and mixes it with Rh$^+$ red blood cells. Coombs reagent is then added. If agglutination occurs, the test is positive, and the mother must be producing anti-Rh antibodies. The *direct*

Coombs test mixes the baby's red blood cells with Coombs reagent. If agglutination occurs, the test is positive, and the baby's red blood cells must have been attacked by its mother's anti-Rh antibodies.

30. In a paternity case, the mother has A$^+$ blood and her son has type O$^-$ blood. If the husband has type B$^+$ blood, which of the following is true?

    A. The husband could be the father.
    B. The husband could not be the father.
    C. The husband could not be the father of an O$^-$ son, but could be the father of an O$^-$ daughter.
    D. The husband is definitely the father.

31. A couple decide to have a child. If the father's genotype is AO and the mother has type B blood of unknown genotype, which of the following are possible blood types for their child?

    I. A
    II. B
    III. AB
    IV. O

    A. I and II only
    B. I, II, and III only
    C. I, II, and IV only
    D. I, II, III, and IV

32. A new virus has been discovered that evades detection by the immune system of only those individuals with type A or type AB blood. Which of the following best accounts for this observation?

    A. The viral antigens resemble the A antigen.
    B. The viral antigens resemble the B antigen.
    C. The viral antigens are Rh$^+$.
    D. The viral antigens are too small to elicit an immune response.

33. If a man with type AB blood needed a transfusion of red blood cells, which of the following individuals could safely donate blood?

    A. A man with type A blood
    B. A man with the genotype BO
    C. A woman with the genotype AB
    D. All four blood types are equally safe

34. How might one most practically assess the risk of *erythroblastosis fetalis* in pregnant women?

    A. Test all women for the presence of anti-Rh antibodies.
    B. Test all fetuses for the presence of the Rh antigen within the first trimester of pregnancy.
    C. Test only Rh⁻ mothers for the presence of anti-Rh antibodies.
    D. Test all mothers of Rh⁺ children for the presence of anti-Rh antibodies.

35. Based on information in the passage, what does the reaction below represent?

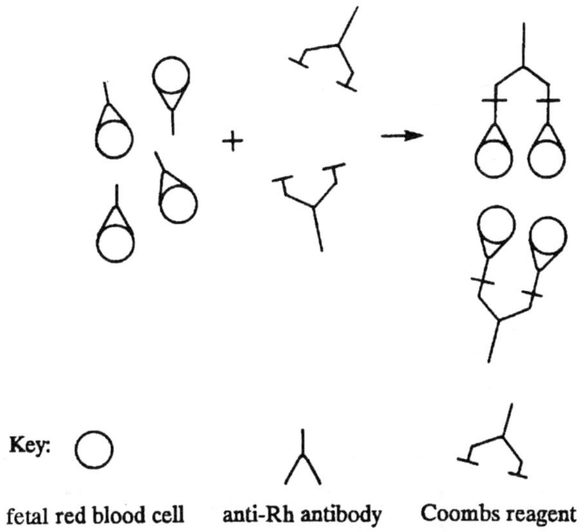

Key:

fetal red blood cell   anti-Rh antibody   Coombs reagent

Figure 3

    A. Negative direct Coombs test
    B. Positive direct Coombs test
    C. Positive indirect Coombs test
    D. Negative indirect Coombs test

36. A woman who has never been pregnant has type B⁻ blood. Which of the following antibodies would you expect to find in her serum?

    A. Anti-B antibody
    B. Anti-A antibody
    C. Anti-Rh antibody
    D. Both anti-A and anti-Rh antibodies

37. A medical student suggested giving Rh⁻ mothers of Rh⁺ fetuses a specific exogenous substance prior to delivery to prevent an immune response. Which of the following substances would likely be the safest and most effective?

    A. Rh antigen
    B. An immunosuppressive drug
    C. Anti-Rh antibody
    D. Iron pills

Questions 38 through 40 are NOT based on a descriptive passage.

38. A certain chemical is found to inhibit the synthesis of all steroids. The synthesis of which of the following hormones would NOT be affected when a dose of this chemical is administered to a laboratory rat?

    A. Cortisol
    B. Aldosterone
    C. Epinephrine
    D. Testosterone

39. A biochemist grows two cultures of yeast—one aerobically and the other anaerobically—and measures the amount of ATP produced by each culture. He finds that the aerobically grown yeast produce about 18 times as much ATP as the anaerobically grown yeast. These observations are consistent with the fact that in the aerobically grown yeast:

    A. oxygen is converted into ATP.
    B. oxygen is necessary to convert glucose into pyruvate.
    C. oxygen is the final electron acceptor of the respiratory chain.
    D. oxygen is necessary for the reduction of pyruvate into lactate.

40. Growth hormone decreases the sensitivity of cellular receptors to insulin. Therefore, a patient with *acromegaly,* which is caused by the oversecretion of growth hormone, would be expected to have:

    A. a low blood glucose concentration.
    B. a high blood glucose concentration.
    C. a decreased urine volume.
    D. a decreased cardiac output.

Passage VI (Questions 41–45)

Although individual organisms have only two alleles for any given trait, it is possible for a trait to have more than two alleles coding for it. This phenomenon is know as multiple alleles. Multiple alleles are created when a single gene undergoes several distinct mutations. These alleles may have different dominance relationships with one another; for example, there are three alleles coding for the human blood groups, the IA, IB, and I alleles. Both the IA and IB alleles are dominant to the I allele, but IA and IB are codominant to each other.

A multiple-allele system has recently been discovered in the determination of hair coloring in a species of wild rat. The rats are found to have one of three colors: brown, red, or white. Let B = the gene for brown hair; b = the gene for red hair; and w = the gene for white hair. The results from nine experimental crosses are shown below. The males and females in Crosses 1, 2, and 3 are all homozygous for hair color.

Table 2

Cross	Male	Female	Offspring
1	brown	red	all brown
2	brown	white	all brown
3	red	white	all red
4	brown	brown	3 brown : 1 red
5	brown	brown	all brwon
6	red	red	all red
7	red	red	3 red : 1 white
8	brown	red	2 brown : 1 red : 1 white
9	brown	red	1 brown : 1 red

41. Based on the experimental results, what is the genotype of the male in Cross 6?

    A. bw
    B. bb
    C. bw or bb
    D. Bb or bw

42. If a large number of brown offspring from Cross 8 are mated with each other, what is the expected percentage of white offspring?

    A. 6.25%
    B. 8.33%
    C. 12.5%
    D. 25%

43. Based on the experimental results, what is the genotype of the female in Cross 5?

    A. Bb
    B. BB or Bb
    C. BB or Bw
    D. BB, Bb, or Bw

44. A white male is crossed with a heterozygous red female from Cross 9. What is the expected ratio of red to white offspring?

    A. 3:1
    B. 1:3
    C. 1:1
    D. 2:1

45. If it were discovered that the alleles for red and white hair were actually incompletely dominant and produced a pink hair color in rats with one copy of each allele, what would be the expected phenotypic ratio in a cross between a Bb male and a pink female?

    A. 2 brown : 1 red : 1 white
    B. 2 brown : 1 red : 1 pink
    C. 1 brown : 2 white : 1 pink
    D. 1 brown : 1 white

Questions 46 through 49 are NOT based on a descriptive passage.

46. A certain drug inhibits ribosomal RNA synthesis. Which of the following eukaryotic organelles would be most affected by the administration of this drug?

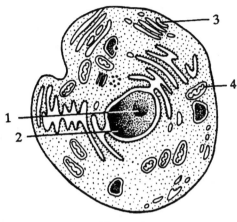

Figure 4

    A. 1
    B. 2
    C. 3
    D. 4

47. Exocrine secretions of the pancreas:

    A. raise blood glucose levels.
    B. lower blood glucose levels.
    C. regulate metabolic rate.
    D. aid in protein and fat digestion.

48. Destroying the cerebellum of a cat would cause significant impairment of normal:

    A. urine formation.
    B. sense of smell.
    C. coordinated movement.
    D. thermoregulation.

49. A cell with a high intracellular K^+ concentration, whose plasma membrane is impermeable to K^+, is placed in an ATP-rich medium with a low K^+ concentration. After several minutes, it is determined that the extracellular concentrations of both K^+ and ATP have decreased, while the intracellular K^+ concentration has increased. What is the most likely explanation for this phenomenon?

   A. The K^+ passively diffused from the medium into the cell.
   B. The K^+ entered the cell by way of facilitated transport.
   C. The ATP formed a temporary lipid-soluble complex with the K^+, thus enabling the potassium to enter the cell.
   D. The K^+ entered the cell by way of active transport.

Passage VII (Questions 50–55)

A powerful recombination mechanism present in both prokaryotic and eukaryotic genomes is the ability of certain DNA sequences to move from one random site to another, either within or between chromosomes. These mobile DNA sequences are called *transposons*.

Transposons are normal components of any organisms' chromosomes. Transposons can be autonomous units, coding for proteins such as the enzyme transposase. Transposons must bind to the transposon itself to sponsor transposition—the insertion of the transposon into a new random target site on the host DNA. Prior to transposon insertion, the target site consists of a unique host DNA sequence. Upon insertion, this sequence is duplicated and flanks the transposon at both ends.

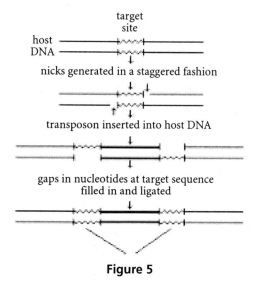

Figure 5

The transposon Tn10 contains two promoters in opposite orientation that lie near the outside orientation that lie near the outside boundary of the transposon (see Figure 6). The stronger promoter, P_{out}, sponsors transcription of the adjacent host DNA. The weaker promoter P_{in}, is the start site from which the gene that codes for transposase is transcribed. There is a 36-base–pair overlap in the transcripts from P_{in} and P_{out}.

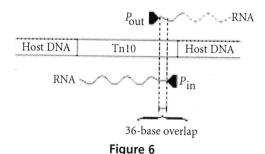

P_{out} ▶ ··········RNA

Host DNA | Tn10 | Host DNA

RNA ~~~~~~~~ ◀ P_{in}

36-base overlap

Figure 6

The effects of *dam methylation* provide an important system for regulating the frequency of Tn10 transposition. Methylation is a common covalent modification to DNA present in most organisms. Since DNA replication is semi-conservative, methylation is initially found only on the original strand. (Newly replicated DNA is thus described as being hemi-methylated.) Following replication, the dam system methylates the new strand. However, the initial absence of methylated sequences on the new DNA strand activates Tn10 by a combination of increasing the transcription of the transposase gene and enhancing the binding of transposase to Tn10.

Multicopy inhibition is another regulatory mechanism employed by Tn10. OUT RNA (mRNA transcribed from P_{out}) is present at 100 times the level of IN RNA (mRNA transcribed from P_{in}). Since OUT RNA base pairs with IN RNA, the excess OUT RNA ensures that IN RNA is bound rapidly before a ribosome can attach.

50. Multicopy inhibition is a regulatory mechanism that limits the expression of the transposase gene by blocking:

    A. replication
    B. transcription
    C. translation
    D. post-transcriptional modification

51. A researcher studying transposition discovered a new strain of cells that have a 1,000-fold greater rate of transposition than other strains. Which of the following explanations would best account for this observation?

    A. The new strain lacks the dam methylation system.
    B. The new strain transcribes 500 copies of OUT RNA per copy of IN RNA.
    C. Only one copy of the target sequences is present after insertion of the transpson into the new strain's DNA.
    D. The overlap between the transcripts from P_{in} and P_{out} in the new strain is twice as many base pairs as the other strains.

52. It can be inferred from the passage that, aside from being involved in the regulation of transposon activity, the presence of hemi-methylated DNA also allows cells to:

    A. divide more rapidly.
    B. distinguish between old and new strands of DNA.
    C. incorporate free methyl groups into other biological molecules, such as amino acids and cofactors.
    D. transport proteins into and out of the nucleus.

53. If a healthy bacterial cell containing Tn10 undergoes cell division and then dies, could the transposon have been responsible for the cell's death?

    A. Yes, because the transposon may have inserted itself into a vital gene, thereby inactivating the gene product.
    B. Yes, because the P_{in} transcript may have inhibited the P_{out} transcript through multicopy inhibition, thereby preventing the production of cellular proteins.
    C. No, because transposons are found in almost all organisms and therefore cannot be lethal if so many genomes possess them.
    D. No, because upon insertion of a transposon the target sequence is duplicated and flanks the transposon at both ends.

54. If Tn10 contained the gene for resistance to the antibiotic ampicillin, which of the following procedures would be the most effective method for identifying *E. coli* cells containing the transposon?

    A. Use differential centrifugation to separate cells on the basis of weight.
    B. Hybridize cells with radio-labeled antibodies specific for *E. coli*.
    C. Use staining and microscopy techniques to determine if cells contain polysaccharide molecules in their cell walls.
    D. Incubate cells on agar plates containing ampicillin.

55. If the first 7 bases of Tn10 IN RNA were 5'-AUAUGCC-3', then what would be the base sequence of the OUT RNA that base pairs with it?

    A. 5'-UAUACGG-3'
    B. 5'-TATACGG-3'
    C. 5'-GGCAUAU-3'
    D. 5'-GGCTUTU-3'

Passage VIII (Questions 56–60)

An embryologist wanted to study the molecular dynamics of the early stages of mammalian development. She chose to examine sea urchin eggs because their early embryonic development is similar to that observed in mammals. Furthermore, mature, unfertilized sea urchin eggs can be divided in two—half with a nucleus, half without—and it is relatively easy to induce "development" in both of these eggs without fertilization. In fact, unfertilized sea urchin eggs—both whole and half—can be *artificially activated* by exposure to a hypertonic solution.

A sample containing mature, unfertilized whole sea urchin eggs, unfertilized half-eggs without a nucleus, and unfertilized half-eggs with a nucleus, was exposed to a hypertonic solution. Protein synthesis was measured following the artificial activation. The embryologist found that, on a gram-for-gram basis, the initial pattern of protein synthesis was comparable in all three egg types (see Figure 7).

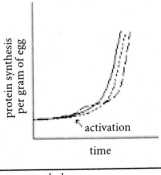

Figure 7

KAPLAN
Test Prep and Admissions

56. Which of the following statements best explains why, on a gram-for-gram basis, the amount of protein synthesized was comparable in all three egg types?

   A. The eggs synthesized the same amount of mRNA, on a gram-for-gram basis, after activation.
   B. The eggs synthesized the same amount of amino acid, on a gram-for-gram basis, during the course of maturation.
   C. The eggs synthesized the same amount of m RNA, on a gram-for-gram basis, during the course of maturation.
   D. The eggs synthesized the same amount of rRNA, on a gram-for-gram basis, after activation.

57. Which of the following graphs would best represent protein synthesis by each of the three egg types if the experiment had run for a longer period of time?

58. Eggs have species-specific receptors on their outer surfaces in which sperm must attach if fertilization is to occur. Would a drug that binds irreversibly to these receptors be an effective contraceptive in mammals?

   A. Yes, because if the receptors are blocked, the egg will atrophy.
   B. Yes, because if the receptors are blocked, sperm will not be able to penetrate the egg.
   C. No, because the egg can be artificially activated by hypertonic solutions of changes in pH.
   D. No, because the sperm nucleus fuses with the egg nucleus prior to receptor attachment.

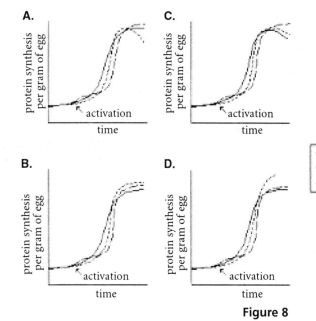

Figure 8

——— whole egg
– – – half egg with a nucleus
------- half egg without a nucleus

59. The half-egg with a nucleus has:

    A. the same number of chromosomes as the whole egg and twice the number of chromosomes as an autosomal cell.
    B. the same number of chromosomes as the whole egg and half the number of chromosomes as an autosomal cell.
    C. the same number of chromosomes as the half-egg without a nucleus and the number of chromosomes as an autosomal cell.
    D. the same number of chromosomes as the half-egg without a nucleus and twice the number of chromosomes as an autosomal cell.

60. In animals that develop externally, the yolk can account for 95% of total egg volume, whereas in mammals, the yolk constitutes less than 5% of total egg volume. Which of the following best accounts for this observation?

    A. Eggs that develop externally must be larger than those that develop internally in order to protect the embryo from predators.
    B. Animal species that develop externally are typically larger than species that develop internally.
    C. Embryos that develop externally generally have a shorter gestation period than embryos that develop internally.
    D. Embryos that develop externally obtain nourishment from nutrients within the egg, while embryos that develop internally obtain most of their nourishment from their mothers.

Questions 61 through 63 are NOT based on a descriptive passage.

61. The nucleus of a tadpole intestinal cell is removed and transplanted into a denucleated zygote. The frog develops normally after the transplant. The experimental results suggest that:

    A. cell differentiation is controlled by irreversibly repressing genes not needed by the cell.
    B. cell differentiation is controlled by selectively repressing genes not needed by the cell.
    C. The cytoplasm of the zygote contains all of the information needed for normal adult development in the form of RNA.
    D. the ribosomes found in the nucleus of the zygote are the same as those found in an adult frog.

62. If albumin, a plasma protein, is infused into arterial blood, which of the following changes would you expect to find in the capillaries?

    A. Increased permeability to albumin
    B. Decreased permeability to albumin
    C. Increased movement of water from the capillaries to the interstitial fluid
    D. Decreased movement of water from the capillaries to the interstitial fluid

63. A 10 year-old boy goes to a physician complaining of polyuria (increased urinary volume and frequency) and weight loss. A blood test reveals elevated blood glucose. What would further tests most likely reveal?

    A. Low insulin
    B. Low ACTH
    C. Low glucagon
    D. Low cortisol

KAPLAN

Test Prep and Admissions

Passage IX (Questions 64–67)

The three-dimensional configuration of an *allosteric protein* is altered in response to the binding of substrate to an active site (or any substrate-binding site) on the molecule. These conformational changes can affect the binding capacity of the other active sites on the molecule. The following two models have been proposed to account for the properties of allosteric multi-subunit proteins with an active site on each subunit.

The *concerted model* postulates that the subunits of an allosteric molecule can exist in either of two interconvertable forms: the inactive T form (Tense), which has low substrate-binding affinity, and the active R form (Relaxed), which has high substrate-binding affinity. In the absence of substrate, nearly all of the molecules are in the T form; the addition of substrate shifts this conformational equilibrium toward the R form. However, all the subunits of a particular molecule must be in the same conformational state at any given time. Hence, the proportion of molecules in the R form increases as more substrate is added, and so substrate binding is cooperative.

In the *sequential model*, the individual subunits of an allosteric molecule exist in either the R form or the T form, but equilibrium between the two forms is not assumed. Rather, the binding of substrate changes only the shape of the subunit to which it is bound; the conformation of the other subunits in the molecule are only negligibly altered. However, the conformational change elicited by the binding of substrate to one subunit can increase or decrease the substrate-binding affinity of the other subunits in the molecule. If, in a protein with two subunits, the RT form has a greater substrate-binding affinity than the TT form, then substrate binding will be cooperative.

64. For an allosteric enzyme that consists of two identical subunits, the existence of which of the following complexes would be consistent with the sequential model but NOT the concerted model?

   A. RR
   B. TT
   C. RRTT
   D. RT

65. The diagram below illustrates the results of an experiment using the allosteric enzyme ATCase and its substrate analog, succinate. Based on these results, which model best accounts for the allosteric properties of ATCase? (Note: f_R = the fraction of ATCase active sites in the R form; Y = the fraction of ATCase active sites with succinate bound.)

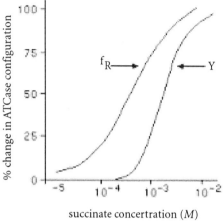

Figure 9

   A. The concerted model, because the Y curve shows cooperative binding.
   B. The concerted model, because the fraction of active sites in the R form (f_R) is greater than the fraction of active sites with substrate bound (Y).
   C. The sequential model, because the fraction of active sites in the R form (f_R) is equal to the fraction of active sites with substrate bound (Y).
   D. The sequential model, because the Y curve has a higher succinate concentration than the f_R curve at any given value on the Y-axis.

66. Which of the following experimental observations would most support the sequential model but NOT the concerted model?

    A. After the addition of an allosteric inhibitor, the conformation equilibrium shifts toward the T form.
    B. After the addition of a saturating concentration of substrate, all subunits have the same conformation.
    C. After a limiting concentration of substrate is bound, the molecule's substrate-binding affinity decreases.
    D. Following denaturation with 8*M* urea, the molecule is unable to bind substrate.

67. The cooperative binding of O_2 to hemoglobin conforms to the concerted model. Therefore, in which of the following structures would you expect the ratio of the R form to the T form to be the greatest?

    A. Right ventricle
    B. Left ventricle
    C. Pulmonary artery
    D. Coronary vein

Passage X (Questions 68–72)

Gout describes a group of diseases characterized by acute and chronic arthritis and severe crippling of the joints. Gout can be caused by an X-linked recessive trait that produces a partial deficiency of the enzyme *hypoxanthine guanine phosphoribosyttransferase* (HGPRT). HGPRT deficiency results in an overproduction of uric acid and accumulation of monosodium urate crystals in the joints and nephrons. Precipitation of urate crystals is believed to be partially responsible for gout.

Uric acid is produced from the breakdown of purines (see Figure 10). Uric acid is deprotonated to urate when the pH is greater than or equal to 5.4. Since most physiological environments are more alkaline than pH 5.4, urate is the predominant form *in vivo*. Xanthine oxidase is the enzyme that catalyzes the conversion of hypoxanthine into xanthine, as well as the conversion of xanthine into urate. Drug therapy is required to reduce the elevated concentration of urate associated with gout. One drug used is *allopurinol*, which is an analog of hypoxanthine and an inhibitor of xanthine oxidase.

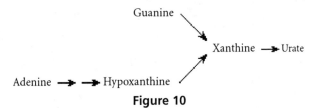

Figure 10

Normally, the serum level of urate is near its solubility limit and approaches *hyperuricemic* conditions, which can result in gout. A high urate concentration is actually beneficial to humans because of its antioxidant property. Urate blocks reactions involving superoxide anions, free radicals, and oxygenated heme intermediates in high valence states. By reducing the rate of attack by these highly reactive species on DNA molecules, urate decreases the rate of mutation.

68. In some patients gout can be controlled with dietary restrictions. Suppression of which of the following metabolic processes would most alleviate hyperuricemia?

    A. Fatty acid catabolism
    B. Nucleotide degradation
    C. Amino acid synthesis
    D. Cholesterol synthesis

69. Which of the following best describes the mechanism of allopurinol action in the treatment of gout?

    A. Allopurinol binds to hypoxanthine, thereby inhibiting its interactions with xanthine oxidase.
    B. Allopurinol binds to xanthine oxidase, thereby promoting the conversion of hypoxanthine to xanthine.
    C. Allopurinol competes with hypoxanthine for xanthine oxidase's active site, thereby blocking the conversion of adenine to urate.
    D. Allopurinol binds to xanthine oxidase, thereby blocking the conversion of adenine to hypoxanthine.

70. If a normal father and a mother with a deficient HGPRT enzyme have a daughter, what is the probability that this child will inherit an HGPRT deficiency?

    A. 0%
    B. 25%
    C. 50%
    D. 75%

71. Inhibition of urate production is just one way to treat gout. The drug colchicine reduces the inflammation of gout by inhibiting granulocyte migration. Based on this information, where is the most likely site of colchicine action?

    A. Microtubules
    B. Ribosomes
    C. Nucleolus
    D. Golgi apparatus

72. A patient is observed to have hypouricemia. In comparison to a normal individual, he would most likely be at a greater risk for:

    A. bacterial infection.
    B. arthritis.
    C. gout.
    D. cancer.

Questions 73 and 74 are NOT based on a descriptive passage.

73. If two individuals with the genotype AaBbCc mate, what is the probability that they will produce offspring with the genotype AABBCC?

    A. 1/64
    B. 1/16
    C. 1/9
    D. 1/3

74. After being exposed to cyanide gas, a culture of aerobic bacteria are infected by a strain of bacteriophage. The culture was later inspected but no signs of viral replication were found. Which of the following explains the finding?

    A. Cyanide binds to viral nucleic acid.
    B. Cyanide destroys bacteriophage binding sites located on the bacterium cell wall.
    C. Cyanide inhibits aerobic ATP formation.
    D. Cyanide denatures bacteriophage enzymes.

MCAT Biological Sciences Test: Answers and Explanations

Answer Key

1. A (See Concept 42.7, pp. 892–895)
2. C (See Concept 42.7, pp. 892–895)
3. C (See Concept 42.7, pp. 892–895)
4. D (See Concept 42.7, pp. 892–895)
5. D (See p. 83)
6. D
7. A (See Concept 42.7, pp. 892–895)
8. C (See Concept 44.2)
9. D
10. A (See Concept 44.4)
11. D (See Concept 44.5, pp. 936–938)
12. C
13. D
14. A
15. B (See Concept 48.5, pp. 1026–1028)
16. C (See Concept 46.3)
17. B (See Concept 46.4, pp. 973–975)
18. A (See Concept 14.4)
19. B (See Concept 14.4)
20. C (See Concept 14.4)
21. C (See Concept 14.4)
22. D (See Concept 14.4)
23. D (See Concept 12.2, p. 221; Concept 15.5, p. 289)
24. A (See Concept 17.7)
25. D
26. A (See Concept 9.5)
27. B (See Concept 15.5, pp. 289–289; Concept 26.4)
28. C (See p. 35)
29. B (See Concept 26.4)
30. A (See Concept 14.3, p. 262; Concept 43.4)
31. D (See Concept 14.3, p. 262; Concept 43.4)
32. A (See Concept 14.3, p. 262; Concept 43.4)
33. D (See Concept 14.3, p. 262; Concept 43.4)
34. C
35. B
36. B (See Concept 14.3, p. 262; Concept 43.4)
37. C (See Concept 14.3, p. 262; Concept 43.4)
38. C (See Concept 45.3)
39. C (See Concept 9.5)
40. B (See Concept 45.3)
41. C (See Concept 14.3, p. 262)
42. A (See Concept 14.2)
43. D (See Concept 14.3, p. 262)
44. C
45. B (See Concept 14.3, pp. 260–261)
46. A (See Concept 6.3)
47. D (See Concept 41.4, pp. 858–861)
48. C (See Concept 48.5, p. 1030)
49. D (See Concepts 7.3 and 7.4)
50. C (See Concept 18.3, pp. 351–352; Concept 19.4, pp. 375–376)
51. A (See Concept 18.3, pp. 351–352; Concept 19.4, pp. 375–376)
52. B
53. A (See Concept 18.3, pp. 351–352; Concept 19.4, pp. 375–376)
54. D (See Concept 18.3, pp. 351–352; Concept 19.4, pp. 375–376)
55. C
56. C (See Concept 21.1; Concept 47.1)
57. A
58. B (See Concept 47.1, pp. 988–991)
59. B
60. D (See Concept 47.1, pp. 992–1001)
61. B
62. D (See Concept 7.3)
63. A (See Concept 45.4, pp. 955–956)
64. D (See Concept 8.5)
65. B (See Concept 8.5)
66. C
67. B (See Concept 42.7)
68. B
69. C
70. A (See Concept 15.3)
71. A
72. D
73. A (See Concept 14.2)
74. C

EXPLANATIONS

Passage I (Questions 1–7)

1. A The key to answering this question lies in knowing that at high altitudes, atmospheric pressure is low, meaning that there is less oxygen in the air than at sea level. We're told that the llama has adapted to life at high altitudes by evolving a different type of hemoglobin. Since the partial pressure of oxygen is lower up in the mountains, llama hemoglobin must be able to bind oxygen more readily at low partial pressures of oxygen. This means that for a given value of oxygen pressure on the x-axis of Figure 1, the llama's hemoglobin will be more saturated with oxygen than the horse's hemoglobin, since horses don't typically live in regions of unusually high altitude. In terms of Figure 1, this means that the llama oxygen-dissociation curve will be to the left of the horse's. So if Curve B is the horse curve, then the llama curve most closely resembles Curve A. Thus, choices (B) and (C) are wrong. Curve D is also wrong: Remember, we're told in the passage that Curves A, B, and C are hemoglobin curves, while Curve D is the myoglobin curve.

2. C According to the passage, small mammals have higher metabolic rates and require a greater amount of oxygen per gram of tissue than larger mammals, and as a result, have hemoglobin that dissociates oxygen more readily than the hemoglobin of large mammals. A high metabolic rate implies that there's a lot of aerobic respiration going on. Metabolically active tissue needs lots of oxygen. The benefit of having Hb that easily dissociates oxygen is that when hemoglobin delivers oxygen to metabolically active tissue, it will readily give up its oxygen to the tissue. This means that for a given value of oxygen pressure, mouse hemoglobin will be less saturated with oxygen than elephant hemoglobin, since an elephant is much larger than a mouse and therefore has a much lower metabolic rate. In terms of Figure 1, the mouse Hb curve will be to the right of the elephant Hb curve. So if Curve B represents oxygen dissociation for elephant hemoglobin, then Curve C most closely resembles the curve for mouse Hb. Therefore, choices A and B are wrong. As for choice (D), we're twice told in the passage that Curve D represents oxygen dissociation for myoglobin. Since we're dealing with hemoglobin in this question, Curve D isn't even an option.

3. C Fetuses are 100% dependent on their mothers for all of their nutritional needs—oxygen being one of them. Oxygen is delivered to the fetus by way of diffusion across the placenta. According to the question stem, Curve A most closely resembles the oxygen-dissociation curve for fetal hemoglobin, assuming that Curve B is the curve for adult hemoglobin. This means that at a given oxygen pressure, fetal hemoglobin is more saturated with oxygen than is adult hemoglobin. This implies that fetal hemoglobin has a greater affinity for oxygen than adult hemoglobin has. In fact, at low partial pressures of oxygen, fetal hemoglobin has a 20–30% greater

affinity for oxygen than adult hemoglobin. That is why oxygen binds preferentially to fetal hemoglobin in the capillaries of the placenta. In addition, fetal blood has a 50% higher concentration of hemoglobin than maternal blood, which increases the amount of oxygen that enters fetal circulation.

4. D The sigmoidal shape of the oxygen-dissociation curve for hemoglobin can be explained by the cooperativity among the subunits of the hemoglobin molecule. According to the passage, hemoglobin is composed of four subunits, each with its own heme group. Each heme unit is capable of binding to one molecule of oxygen, and so the entire molecule is capable of binding four molecules of oxygen. The binding of oxygen at the first heme group induces a conformational change in the hemoglobin molecule such that the second heme group's affinity for oxygen increases. Likewise, the binding of oxygen at the second heme group increases the third heme's affinity for oxygen, and the binding of oxygen at the third heme group increases the fourth's affinity for oxygen. Therefore, the partial pressure of oxygen and the % oxygen-saturation of hemoglobin are not linearly proportional. As a consequence of these shifts in oxygen affinity with each binding, the line representing the oxygen-dissociation curve for hemoglobin is not straight, but rather a sigmoidal, or S-shaped, curve. Thus, choice (D) is the right answer. Choice (A) is wrong because when the iron molecule of the heme group binds to oxygen, it is reduced; when the iron releases the oxygen, it is oxidized. However, this neither results in the sigmoidal shape of the curve, nor does it affect it. The concentration of carbon dioxide in the blood, choice (B), is a factor that does affect hemoglobin's affinity for oxygen and therefore affects the positioning of the curve, but it is not responsible for the sigmoidal shape. A high concentration of carbon dioxide in the blood will decrease hemoglobin's affinity for oxygen, and will therefore shift the curve to the right. Choice (C) is also a true statement; myoglobin does have a higher affinity for oxygen than does hemoglobin, as shown in Figure 1. However, this does not affect the shape of the sigmoidal curve, so choice (C) is also incorrect.

5. D The four subunits in Hb are held together by noncovalent interactions. So placing a sample of human adult hemoglobin in an 8M urea solution, which you're told disrupts noncovalent interactions, will cause the subunits to break apart. You're also told in the question stem that the α chains of this sample of hemoglobin were isolated. So you need to figure out what the oxygen dissociation curve of a single peptide chain would look like. From the passage you also know that myoglobin consists of a single polypeptide chain. Therefore, the oxygen-dissociation curve for one polypeptide chain of Hb would be expected to look similar to the curve for myoglobin. In fact, both the individual α chains and the β chains of hemoglobin resemble the tertiary structure of myoglobin. Thus, the curve for the α chain will look like Curve D, so choice (D) is correct. The single chain of Hb will not look like Curves A, B, or C because these curves have a unique shape due to the cooperativity of the four hemoglobin subunits. Since you're now dealing with a single chain (because of that treatment with an $8M$ urea solution), no cooperativity is possible.

6. D You're told that the utilization coefficient is the fraction of the blood that releases its oxygen to tissues under normal conditions, and that under these conditions, the value of the coefficient is approximately 0.25. During strenuous exercise, there is a greater demand for oxygen, especially in skeletal muscle, where oxygen supplies are rapidly depleted during cellular respiration. Under such conditions, one would expect that a greater fraction of the blood would give up its oxygen, and that the utilization coefficient would therefore be some value greater than 0.25. Also, during strenuous exercise, the utilization coefficient has been recorded at values ranging between 0.75 and 0.85, meaning that 75–85 percent of the blood gives up its oxygen in tissue capillaries. Since choice (D) is the only value greater than 0.25, it is the right answer.

7. A Figure out how much oxygen will be bound to myoglobin, if the myoglobin in this sperm whale is completely saturated with oxygen. All of the answer choice values are in moles. To equate the number of moles of oxygen that will bind to myoglobin, you need to figure out how many moles of myoglobin are present in the sperm whale's muscles. From the question stem you know that there are 0.004 moles of myoglobin per kg of muscle and the whale has 1,000 kg of muscle. Multiplying these two numbers together gives you 4 moles of myoglobin, implying that the whale has 4 moles of myoglobin in its muscles. Then, determine how many molecules of oxygen bind to a single molecule of myoglobin. The answer is one, because myoglobin consists of a single polypeptide chain, which contains a single heme

group. Thus, myoglobin binds only one molecule of oxygen. If one molecule of oxygen binds to one molecule of myoglobin, then one mole of oxygen will bind to one mole of myoglobin. Since you know that the whale has 4 moles of myoglobin in its muscle, 4 moles of oxygen will be bound to myoglobin when myoglobin is completely saturated with oxygen.

Passage II (Questions 8–14)

8. C According to the passage, the catabolism of amino acids, which are molecules that make up proteins, produces ammonia. Metabolism can be divided into catabolism and anabolism, and catabolism refers to pathways in which larger molecules are broken down into smaller parts. (If you have trouble remembering the difference between catabolism and anabolism, just associate anabolism with anabolic steroids. Anabolic steroids are used by weight lifters to build muscle, therefore anabolism must refer to the pathways in which smaller molecules are built into larger molecules.) We know from the passage that if the ammonia that is produced as a by-product of amino acid breakdown is not further degraded, it can become toxic, depending on its concentration in the blood. To answer the question, you need to recall that the end product of amino acid degradation is urea, which means that ammonia must feed into the urea cycle. The concentration of ammonia in the blood would become elevated and potentially toxic if one of the enzymes involved in the urea cycle was defective. Thus choice (C) is the correct answer. Choice (A), glycolysis, refers to the catabolism of glucose into pyruvate. Choice (B), fatty acid metabolism, refers to the breakdown of fatty acids into acetyl CoA. Choice (D), nucleic acid degradation, refers to the breakdown of DNA and RNA into nucleotides.

9. D To answer this question you need to recall the site of protein digestion, and the fate of the amino acids produced by this digestion. A brief review of the digestive system as it relates to proteins is in order. Food enters the stomach from the pharynx and esophagus. When protein-containing food reaches the stomach, its presence causes the release of the hormone gastrin, which stimulates the secretion of HCl, and pepsinogen, and muscular contractions of the stomach. HCl initiates the conversion of pepsinogen to its active form, pepsin. Pepsin breaks down proteins into peptides. The peptides then enter the small intestine. Three types of peptide-digesting enzymes are secreted into the small intestine: enterokinase, aminopeptidases, and dipeptidases. The combination of these three enzyme types breaks apart all peptide bonds, producing only single amino acids and dipeptides. The single amino acids and dipeptides are then absorbed into the epithelial cells lining the small intestine by active transport. These molecules all enter the bloodstream by way of the capillaries of the villi. Villi are the fingerlike projections lining the small intestine that serve to increase the absorptive surface area of the intestine. From the bloodstream the amino acids enter individual cells. Inside these cells, the amino acids are used to either build new proteins or other biological molecules, or they are completely catabolized, in which case ammonia is produced. This ammonia then enters the urea cycle, and the urea produced in the process exits the cells and is excreted in the urine. So from this discussion of the digestion of protein, it is clear that ammonia is not produced inside the lumen of the small intestine,which makes choice (D) the correct answer.

10. A Albumin is a protein produced by the liver. It is very important in maintaining the plasma osmolarity of the blood. The glomerulus, which is the network of blood vessels enveloped by the part of the nephron known as Bowman's capsule, has small holes in its endothelial lining called fenestrations. When proteins travel through the glomerulus, they are prevented from entering the nephron due to the size of the fenestrations and the negative electrical charge of glycosylated proteins lining the fenestrations. Therefore, albumin is not normally found in urine because it is prevented from entering the nephron, so choice (A) is the right answer. The presence of protein in a urine sample is typically a sign of renal disease.

Each of the other choices is normally found in appreciable quantity in the urine. Urea, choice (C), is the primary excretory by-product of the urea cycle, while sodium and potassium homeostasis are kept in balance by the regulation of their excretion in urine.

11. D As explained in the passage, diabetes insipidus is a condition in which ADH is ineffective, and as a result, the kidneys reabsorb less water and are unable to concentrate urine. Very dilute urine is the result. So, urine output is increased in a patient with diabetes insipidus, which means that choice (D) is correct. Since water reabsorption is decreased, this means that plasma osmolarity will be increased, so choice (A) is wrong. Choice (C) is wrong because the

presence of glucose in the urine is one of the symptoms and tell-tale signs of diabetes mellitus. Diabetes mellitus is the result of an insulin deficiency; without insulin, glucose is not converted into glycogen, and as a result, some of it gets excreted in the urine because the kidneys are overwhelmed by the excess glucose in the bloodstream. Choice (B) is incorrect because an increased urine osmolarity would mean that there was a greater concentration of dissolved solutes per liter of urine, which is not the case if there is an excess of water being excreted. It should be noted that a diabetes insipidus patient with an intact thirst mechanism and access to water will not become dehydrated. They will drink enough water to replace what is lost in the urine.

12. C You're told that a normal value for urine osmolarity is 285 milli-osmoles per liter of H_2O. Since ADH increases water reabsorption in the kidneys, patients with diabetes insipidus are expected to have a decreased urine osmolarity. And if normal is 285 milli-osmoles per liter of water, then, based on the information in Table 1, you should have concluded that Patient A does not have diabetes insipidus. Patients B, C, and D, have a very low urine osmolarity prior to therapy, indicating that there is something wrong. To answer this question you must have a good understanding of the mechanisms behind both central and nephrogenic diabetes insipidus. According to the passage, central diabetes insipidus is when ADH itself is either deficient in quantity or quality. Therefore, exogenous supplementation of ADH should alleviate the symptoms: that is, the kidneys should be able to concentrate urine, and therefore, urine osmolarity should greatly increase only after ADH is administered. It should not increase after the 24-hour restriction of fluid intake because water is not being reabsorbed. If you look at the results of the four patients, you'll see only Patient C's urine osmolarity increased after ADH was administered. So, choice (C) is the right answer. In the case of nephrogenic diabetes insipidus, it is the tubules themselves that are defective. Exogenous ADH would be ineffective as therapy since the patient's own ADH is sufficient in quality and quantity.

13. D Nephrogenic diabetes insipidus is when the kidney's collecting tubules are unresponsive to ADH. It's not that ADH is insufficiently produced or defective in any way, it's that the tubules do not concentrate urine by becoming more permeable to water when ADH is secreted by the posterior pituitary. Again, as in the previous question, you should have immediately ruled out Patient A because Patient A has a normal urine osmolarity prior to the therapy. A patient with nephrogenic diabetes insipidus would have a decreased urine osmolarity prior to therapy, which would remain low even after the restriction of fluid intake for 24 hours. This is because the tubules won't reabsorb water, regardless of the amount of fluid ingested. Likewise, the administration of ADH will not affect the urine osmolarity because, as said earlier, the patient's ADH is perfectly normal; it's the nephron tubules that are defective. If you look at Table 1, you'll see that only Patient D fits the bill. Patient D is most likely suffering from nephrogenic diabetes insipidus, which means that choice (D) is the right answer.

14. A Fluid restriction is the first step in the attempt to diagnose the pathology, if there is one, behind the inability to concentrate urine. Patient B had a low urine osmolarity prior to the onset of therapy. However, following the restriction of fluid intake, there was a substantial increase in urine osmolarity, indicating that Patient B's dilute urine is a function of the amount of fluid ingested. If a lot of fluid is drunk during the day, the urine formed will be dilute, assuming that the person's kidneys and ADH are both normal. Likewise, if there is very little fluid drunk during the day, the urine formed will be concentrated. Thus, the excessive intake of water seems to be the most likely explanation for the formation of dilute urine in Patient B. In fact, there are individuals who have a psychological condition in which they drink water excessively. This excessive intake causes the output of dilute urine. Simply restricting and observing fluid intake would enable the patient to concentrate urine. Fluid restriction had no effect on either Patient C or Patient D, indicating that they have a true pathology; choices C and D are wrong. If Patient B was dehydrated, then the body's response would be to produce concentrated urine from the start, which was clearly not the case here, since urine osmolarity was very low prior to therapy.

Discrete Questions

15. B The vertebrate nervous system consists of two main parts: the central nervous system (the brain and the spinal cord), and the peripheral nervous system, which is divided into the sensory division and the motor division. The sensory division consists of those receptors and neurons that transmit signals to the central nervous system. The motor division transmits signals from the central nervous system to effectors, and is divided into the somatic nervous system and the autonomic nervous system. The somatic system, choice (C), innervates skeletal muscle, and its nervous pathways are typically under voluntary control. The autonomic nervous system regulates the internal environment by way of involuntary nervous pathways. The autonomic nervous system innervates smooth muscle in blood vessels and the digestive tract, and innervates the heart, the respiratory system, the endocrine system, the excretory system, and the reproductive system. The autonomic system is further divided into the sympathetic division, choice (B), and the parasympathetic division, choice (A). The sympathetic division innervates those pathways that prepare the body for immediate action; this is known as the "fight-or-flight" response. Heart rate and blood pressure increase, blood vessels in the skin vasoconstrict and those in the heart vasodilate, pathways innervating the digestive tract are inhibited, and epinephrine, or adrenaline, is secreted by the adrenal medulla, thereby increasing the conversion of glycogen into glucose. This discussion makes it obvious that the correct answer is choice (B); activation of the sympathetic nervous system is associated with an increased heart rate, blood pressure, and blood glucose concentration. The parasympathetic system innervates nervous pathways that return the body to homeostatic conditions following exertion. Heart rate, blood pressure, and blood glucose concentration all decrease, blood vessels in the skin vasodilate and those in the heart vasoconstrict, and the digestive process is no longer inhibited.

16. C The epididymis, choice (A), is a group of coiled tubes sitting on top of the seminiferous tubules in the male reproductive tract. Sperm are produced in the seminiterous tubules and mature and acquire motility in the epididymis. Sperm are stored in the epididymis until ejaculation. Hence, the epididymis functions in the male reproductive system only; choice (A) is wrong. Choice (B), prostate, is one of the glands associated with the male reproductive tract; the prostate gland secretes an alkaline milky fluid that protects the sperm from the acidic conditions in the female reproductive tract. So, choice (B) also functions only in the male reproductive system. The urethra, choice (C), is a structure found in both men and women. During ejaculation, sperm travels from the epididymis, through the vas deferens, and through the urethra, which opens to the outside from the tip of the penis. The urethra is also directly connected to the bladder. Hence, in males, the urethra functions in both the reproductive and excretory systems. In females, however, the reproductive and excretory systems do not share a common pathway. Sperm enter the vagina and travel up through the cervix, uterus, and fallopian tubes, and urine leaves the body through the urethra; the vagina and the urethra never meet—they are separate openings. So, choice (C) is the right answer. Choice (D), ureter, is the duct connecting the kidney to the bladder. Urine is formed in the kidneys, travels down to the bladder by way of the ureters, and is stored there until it is excreted through the urethra. This process is the same in both sexes.

17. B In gametogenesis in males, diploid cells called spermatogonia undergo mitosis to produce diploid cells called primary spermatocytes. The primary spermatocytes undergo the first round of meiosis to yield secondary spermatocytes, which are haploid. The secondary spermatocytes undergo the second round of meiosis, resulting in four haploid cells called spermatids. The spermatids then mature into sperm—the male gametes. So, choice (A) is incorrect because a spermatogonium is a diploid cell. We've already found the correct answer—choice (B), spermatid, is not a diploid cell, it's haploid. In female gametogenesis, a diploid cell called a primary oocyte undergoes the first meiotic division to yield two haploid cells—a polar body and a secondary oocyte. The secondary oocyte undergoes the second meiotic division to produce two more haploid cells—a mature oocyte, or ovum, and another polar body. So choice (D), primary oocyte, is also incorrect because these are diploid cells. During fertilization, an ovum and a sperm fuse; two haploid cells fuse to form a single diploid cell called a zygote. Thus, choice (C) is also incorrect.

Passage III (Questions 18–22)

18. A Choice (A) states that Albert did not have the gene for hemophilia. This is true; he doesn't have the gene, as indicated by his unshaded square. If he did have the gene then he would have been a hemophiliac, since the gene is X-linked. So choice (A) is the correct answer. Choice (B) states that Queen Victoria had two X chromosomes, each with the hemophilia gene. It's true that Queen Victoria had two X chromo-

somes—all normal females do. However, as indicated by the half shading on her circle in the pedigree, she was a carrier of the gene for hemophilia; that is, she had only one copy of the gene. So choice (B) is wrong. Choice (C) states that neither Albert nor Victoria had the gene for hemophilia. Victoria was a carrier, so choice (C) is wrong, too. Choice (D) states that Albert was a carrier, which is impossible because men cannot be carriers of an X-linked trait. If they have a copy of the gene in question, then they express the trait. Males cannot have two copies of an X-linked trait because they have only one X chromosome.

19. B A male can inherit an X-linked trait only from his mother, since he inherits his one X chromosome from her. Thus, Louis IV, who was normal, did not inherit the gene for hemophilia from his mother—who is not shown on the pedigree. Remember, Louis IV was not a blood relative of the Queen; he was her son-in-law. Choice (A) is incorrect, since Louis IV's son, Frederick, has hemophilia because he inherited the gene from his mother, who was a carrier (although we don't know her name). Besides, whether Louis IV's children were hemophiliacs, carriers, or normal is irrelevant to the discussion of why Louis himself was normal. Choice (C) is incorrect because Albert was not a blood relative of Louis IV, so Albert's genotype is independent of Louis' genotype. As for choice (D), while it is true that only females can be carriers of the hemophilia gene, or of any X-linked gene, for that matter, this does not answer the question.

20. C If you look at the pedigree, you'll see that Beatrice—a member of the F_1 generation—was a carrier of the gene for hemophilia, which means that she had one copy of it on one of her X chromosomes. In reality, Beatrice married a normal male, whose name you're not given, but for the purpose of this question, you're asked to determine the probability that any of her sons would have been hemophiliacs if she had in fact married a hemophiliac. This is basically a cross between a carrier and a hemophiliac. Therefore, Beatrice's genotype is X^hX, and her theoretical husband's genotype is X^hY. In a cross between these two people, 50% of all of their children are expected to be hemophiliacs. But you need to look at the disease in terms of gender: 50% of the daughters are expected to be hemophiliacs; the other 50% will be normal. Likewise, 50% of the sons are expected to be hemophiliacs; the other 50% to be normal. So 50% is correct, which is choice (C).

21. C Victoria Eugenia had four sons, three of whom were hemophiliacs; i.e., 75% of her sons turned out to be hemophiliacs. You know that these sons inherited the disease from their mother, as opposed to their father, Alfonso III of Spain, because their mother was a carrier of the disease, as can be seen on the pedigree, and their father was normal. Seventy-five percent was the actual percent, but the question asks you to determine the theoretical probability that Victoria Eugenia's sons would be hemophiliacs. So, what you have to do is work out the cross between a carrier female and a normal male and look at the results. If you do that, you'll find that theoretically, 50% of Victoria Eugenia's daughters should have been carriers and 50% should have been normal, while 50% of her sons should have been hemophiliacs and 50% should have been normal. So the probability that Victoria Eugenia's sons would have been hemophiliacs is 50%, choice (C).

22. D According to the pedigree, Rupert's father was normal, his mother, Alice, was a carrier of the disease, his maternal grandmother was normal, and his maternal grandfather, Leopold, was a hemophiliac. Knowing this bit of family history allows us to reconstruct the transmission of the gene for hemophilia through these three generations, and also allows us to rule out choices (B) and (C), since Alice was not a hemophiliac and her husband was normal. Leopold, the hemophiliac, must have passed on the hemophilia gene to his daughter Alice, since she inherited one of her X chromosomes from him. Alice's mother was normal, so Alice's other X chromosome was normal. Alice married a normal man, and gave birth to three children, a normal daughter and son, and a hemophiliac son, Rupert. Rupert had hemophilia because he inherited the X chromosome with the hemophilia gene on it from his mother, who, in turn, inherited it from her father. Rupert's hemophilia is a direct result of his maternal grandfather's hemophilia, and therefore choice (D) is the right answer. Choice (A) is wrong because a mutation of the Y chromosome that Rupert inherited from his father would not affect Rupert's inheritance of hemophilia, since the gene for hemophilia is located on the X chromosome, not the Y.

Passage IV (Questions 23–29)

23. D During the G_1 phase, the cell undergoes intense biochemical activity. During the S phase, or synthesis phase, the nuclear DNA replicates. During the G_2 phase, the nuclear DNA condenses and the

structures used during mitosis begin to assemble. During the M phase, or mitotic phase, mitosis occurs. Finally, the nuclear DNA segregates to opposite poles of the cell, the replicated organelles—mitochondria included—also segregate, along with the nuclear DNA, and the cell divides, forming two identical daughter cells. In the passage you're told that mitochondria replicate in a seemingly random pattern, out of phase with both other mitochondria and the cell itself. Remember that mitochondrial DNA must replicate in order for the mitochondrion itself to replicate. Basically, it can be inferred that mitochondrial DNA replicates throughout all of the phases of the cell cycle: G_1, S, G_2, and M. So, choice (D) is correct.

24. A Mitochondria are responsible for cellular energy production—they supply the cell with ATP. Mitochondrial DNA directs the synthesis of mitochondrial proteins, which ultimately play a major role in cell survival. Since mitochondria are so essential to eukaryotic cell life, one would therefore expect replication of its DNA to be highly accurate. Mutations that would cause a dramatic change in its DNA and its ability to produce proteins needed for ATP formation would be lethal to the cell. Since mutations do occur, the most likely type to occur would be one that causes the least damage. A point mutation fits the bill. Point mutations are defined as those in which only nitrogenous base is affected; for example, a cytosine is substituted for an adenine during replication. Point mutations are not usually lethal because of the redundancy of the genetic code; that is, each amino acid is typically coded for by more than one codon. Take the amino acid proline, for example. Proline is coded for by four codons: CCU, CCC, CCA, and CCG. Let's say that the codon is CCU; if there's a point mutation at the third base, no matter which of the remaining three bases is substituted for the uracil, the net product will still be proline. So a point mutation is the type of mutation least likely to affect their productivity. Thus, choice (A) is correct.

As to the other choices, Choice (B), frameshift mutation, is a mutation causing genetic material to be inserted or deleted during DNA replication or transcription. This produces a shift in the reading frame of the mRNA strand being translated, usually leading to the formation of nonsense polypeptides. Changes in protein synthesis would most likely be dangerous for the mitochondria and the cell itself. Lethal mutations, choice (C), are those that would cause the mitochondria to become nonfunctional. In

Choice (D), nondisjunction is the failure of homologous chromosomes to separate during meiosis. First of all, mitochondria have one circular chromosome—there aren't any homologous chromosomes. Second, mitochondria do not undergo meiosis—only specialized eukaryotic cells in sexually reproducing organisms undergo meiosis. So, choice (D) cannot even be a consideration.

25. D   The nuclear genome is comprised of double-helical DNA that codes for mRNA, tRNA, and rRNA. Therefore, choices (A) and (B) are characteristics of the nuclear genome and are therefore incorrect. If a mutation occurred in the nuclear genome that rendered an essential gene nonfunctional, such as an enzyme involved in glycolysis, the organism would die. Thus, choice (C) is also a characteristic of the nuclear genome and an incorrect choice. Although the nuclear genome encodes many products, most of the bases of DNA are non-coding. That is, they are involved in the regulation of gene expression and do not themselves code for any product. So, choice (D) is not consistent with the nuclear genome, so choice (D) is the correct answer. The mitochondrial genome is so small, compared to that of the nucleus, that almost every nitrogen base has to code for a product; the mitochondrial genome doesn't have any DNA to waste!

26. A   To answer this question, you must know how many ATP are formed from the catabolism of one molecule of glucose by an obligate anaerobe. An obligate anaerobe is an organism that must live without oxygen in order to survive. Obligate anaerobes produce ATP via fermentation, which includes both glycolysis and the reactions necessary to regenerate the NAD^+ necessary for glycolysis to continue. Fermentation leads to a net production of 2 ATP; this ATP is generated during glycolysis. Therefore, an obligate anaerobe will produce 2 ATP per molecule of glucose. Aerobic organisms produce a net of 36 ATP, choice (D), per molecule of glucose, as shown in the equation provided in the passage. So, choice (D) is wrong, while choices (B) and (C) are just nonsense.

27. B   From the question stem you know that the mating type of a wild-type strain, which has normal mitochondrial DNA, is crossed with the opposite mating type of a strain that lacks functional mitochondria due to deletions in the mitochondrial genome. "Mating types" is a way of referring to male and female in species that do not technically

have opposite genders, such as algae and yeast. In addition, you're told that the offspring of this cross do not have functional mitochondria either. The offspring have the same deleted mitochondrial genome as the mutant strain. Now all you have to do is find the choice that best accounts for this occurrence. Choice (A) is incorrect because the endosymbiotic theory attempts to explain the derivation of mitochondria in eukaryotic cells, not the inheritance of mitochondria. Choice (B) is correct. Since you're told that the offspring lack mitochondrial functions, this implies that they inherited their mitochondria from the mutant strain mating type. In other words, the mutant strain was the organelle-donating parent—the female—in this cross. Therefore, the non-Mendelian inheritance pattern of mitochondria, as explained in the passage, best accounts for these experimental observations. If the mating type of the wild-type strain had been the organelle-donating parent, all of the offspring would have normal mitochondrial functions. Choice (C) is wrong because the word recombination implies the formation of new gene combinations due to crossing over events that occur during reproduction. If recombinations did occur, you would expect some of the offspring to regain mitochondrial functions, since wild-type mitochondrial DNA would replace the deleted segments of DNA in some offspring. Although choice (D) is a true statement, it does not explain the inheritance patterns observed in this cross.

28. C   According to the question stem, four different human cell cultures were grown in a medium containing radioactive adenine. The first thing that you should be thinking about is DNA replication; while these cells are replicating they're going to incorporate this radioactive adenine into all of their DNA. This includes chromosomal DNA, as well as mitochondrial DNA. Mitochondria replicate independently of their cells. Since all autosomal human cells have the same amount of DNA in their nuclei, the only difference in radioactivity will be the amount that was incorporated into the mitochondrial DNA. This is why the cells' mitochondria were isolated via centrifugation, and the radiation from each sample was measure using a scintillation counter. The cells with the greatest number of mitochondria will have the highest radioactive count when their mitochondria are separated. So, you need to determine which of the four cell types would have the greatest number of mitochondria. This number is dependent on the energy needs of the tissue. Given the choices—erythrocytes, epidermal cells, skeletal muscle cells, and intestinal cells—you should know that the correct answer is choice (C), skeletal muscle cells. Muscle cells need a lot of energy in order to contract. ATP is required every time a molecule of myosin binds to actin in the sarcomeres. In general, muscle cells have a higher content of mitochondria than do any other type of autosomal cell. Erythrocytes, choice (A), do not even contain any mitochondria. Choice (B), epidermal cells, do not have any special energy requirements. The same thing applies to intestinal cells, choice (D).

29. B In the passage you're told that many scientists believe that mitochondria were once independent unicellular entities, possibly prokaryotic in origin, which formed a symbiotic relationship with eukaryotic cells that was mutually beneficial to both parties. This theory is known as the endosymbiotic hypothesis. If mitochondria are believed to have been prokaryotic in origin, then any findings that reveal similarities between mitochondria and bacteria would support the hypothesis. The question asks you to find an answer choice that does not support this hypothesis. Choice (A) says that mitochondrial DNA is circular and not enclosed by a nuclear membrane, which is, in fact, true. Bacteria have a similar structure in that they have a single circular chromosome located in a region of the cell known as the nucleoid, which is not bound by a membrane. So choice (A) is incorrect because it supports the endosymbiotic hypothesis. Choice (B) states that mitochondrial ribosomes more closely resemble those found in eukaryotes than those found in prokaryotes. This does not support the hypothesis—any similarity to eukaryotic cells doesn't support it. So choice (B) appears to be the correct answer. Choice (C) says that there are many present-day bacteria that have symbiotic relationships with eukaryotic cells. This does support the concept that the mitochondrial ancestor could have lived within a eukaryotic cell in a mutualistic relationship. So, choice (C) is wrong because it supports the hypothesis. Choice (D) says that mitochondrial DNA codes for its own ribosomal RNA. This provides evidence that at some point, mitochondria existed as free entities, capable of directing their own protein synthesis, cell division, and all other cellular activities associated with independently living organisms; choice (D) is also incorrect. By the way, the information contained in choice (B) is not true; mitochondrial ribosomes actually do resemble

prokaryotic ribosomes, thereby providing further support for the endosymbiotic hypothesis. However, if it were true, it would not support the hypothesis.

Passage V (Questions 30–37)

30. A Let's determine as much as we can about the blood types of the parents and the son. We know that the mother is blood type A$^+$, and therefore one of her alleles codes for the A antigen. We don't know whether the other allele for that locus is an O or an A. The mother is also Rh$^-$, and again, similar reasoning applies here. We know that she has at least one Rh allele that codes for the Rh antigen, which makes her Rh$^+$, since the Rh allele is dominant. The second allele at that locus might be either Rh$^+$ or Rh$^-$. The son is O negative, so we know that his genotype for the ABO blood groups is OO. Likewise, with respect to the Rh locus, he's doubly negative. If he had just one Rh allele he would be Rh$^+$, since the allele is dominant. So the son's genotype must be OO Rh$^-$, Rh$^-$. Working backwards, we can deduce that his mother's genotype must be AO Rh$^+$, Rh$^-$, since the son inherited one allele per locus from his mother. The husband is type B positive, so we know that he has to have at least one B allele and one Rh allele. In the paternity case, the question to be addressed is, is it possible for the husband to be the father of this boy, knowing what we know about the mother? The answer is yes, it's possible. We know that the mother's genotype is AO Rh$^+$, Rh$^-$; now suppose that the husband's blood type is BO Rh$^+$, Rh$^-$. If this were the case, then the son could have inherited one O allele from each parent and one Rh$^-$ allele from each parent. Therefore, it is possible for these two people to conceive an O negative child. Thus, the correct answer is choice (A); the husband could have been the father.

31. D The A and B alleles are codominant to each other, and the O allele is recessive. Codominance means that both of the alleles are phenotypically expressed. So, when a person has both the A and the B alleles, that person is said to have type AB blood and expresses the properties ascribed to both alleles—that is, his or her red blood cells have both the A and B antigens on their surface. During sexual reproduction, the mother and father each donate one allele to their offspring. In this case we know that the father's genotype is AO; he can donate either an

A allele or an O allele to his child. We don't, however, know the mother's genotype; we only know that her phenotype is type B blood. This means that her genotype is either BO or BB—we simply don't know which one it actually is. Let's first assume that the mother has the genotype BB and must therefore donate a B allele to the child. In this situation, if the father donates an A, the child's genotype and phenotype will be AB. If the father donates an O allele, the child's genotype will be BO and the phenotype will be type B. Therefore, statements II and III are correct. Since III is correct, you can rule out choices (A) and (C) because they don't contain it. Now let's assume the mother's genotype to be BO. This means that she can donate an O allele to the child. In this case, if the father donates an A allele, the child's genotype will be AO and the phenotype will be type A. This means that statement I is also correct, but this doesn't help you decide between choices B and D, because they both contain I. (Also, I must be correct because it appears in both of these remaining choices.) So what it comes down to is whether or not this child could have type O blood. If the father donates an O allele, the child's genotype will be OO and the phenotype will be type O. This means that statement IV is also correct; all four blood types are possibilities. Thus, choice (D) is the correct answer.

32. A From the question stem you know that this virus avoids detection by the immune system of only those individuals who have type A or AB blood. From this fact you can conclude that the virus must not evade the immune system of people with B or O blood types. The correct answer will differentiate between these two groups. From this piece of information, you can eliminate choice (D). If the viral antigens are too small to elicit an immune response, the virus would evade all immune systems, not just those of individuals with A or AB blood types. Choice (C) can also be eliminated since the presence of the Rh factor is independent of ABO blood type. So now we need to decide between choices A and B. How do type A and type AB blood differ from type B and type O blood? Individuals with type A and AB blood express the A antigen on the surface of their red blood cells. These individuals will therefore not produce antibodies to the A antigen. On the other hand, individuals with type B and O blood do not express the A antigen, and will produce antibodies to the A antigen. If the virus' antigen mimicked the A antigen, people who normally express the A antigen—that is, people with

type A or type AB blood—would not recognize the virus as foreign. No immune response would be elicited, and the virus would be able to persist in these people. Thus, choice (A) is the correct answer. If the viral antigens mimicked the B antigen, as in choice (B), the virus would evade detection by the immune system in those individuals who normally expressed the B antigen, type B and type AB people, but would be detected by the immune system of those individuals who do not normally express the B antigen, type A and type O people. Thus, choice (D) is also incorrect.

33. D The thinking process behind this question is similar to the one used to answer the previous question. A person with type AB blood expresses both the A and B antigens on his red blood cells, which implies that his blood does not contain any anti-A or anti-B antibodies. Since the recipient's blood does not contain anti-A antibodies or anti-B antibodies, any blood type can be safely transfused, regardless of the A and B antigens found in the donor's blood. Be aware that there are other blood antigens typically present that could cause problems during transfusions, but this is beyond the scope of this question. Also recognize that the gender of the person donating the blood is in no way relevant.

34. C Keep in mind that you're asked to choose the most practical screening program. Rh incompatibility exists when an Rh⁻ mother, who has been sensitized to the Rh antigen by a previous pregnancy, is pregnant with another Rh⁺ baby. The risk exists that the mother, as a result of this previous exposure to the Rh antigen, has produced antibodies that will now attack the fetus' red blood cells, which are Rh⁺. So, the first step in making the assessment of erythroblastosis fetalis risk is to narrow down the group being assessed. The mother must definitely be Rh⁻ for there to be any risk at all, which means that testing all women, as in choice (A), is extremely impractical. So choice (A) is incorrect. Choice (B) is incorrect because it tests all fetuses for the presence of the Rh antigen. This is useless unless the mother is known to be Rh⁻. The most practical test is the one that singles out Rh⁻ mothers for testing. So, choice (B) is also incorrect. Choice (D) is not practical because all mothers of Rh⁺ children, regardless of their own blood type, are tested. Both Rh⁺ and Rh⁻ mothers would be tested, which is unnecessary and expensive. So, choice (C) is the correct answer; test

Rh⁻ mothers to determine whether or not they are producing anti-Rh antibodies. If the test is positive, then there is a risk of erythroblastosis fetalis for an Rh⁺ fetus.

35. B The Coombs tests are the screening procedures used to assess whether an Rh incompatibility reaction can or has occurred. The direct Coombs test will identify whether the baby's red blood cells have in fact been attacked by the mother's anti-Rh antibodies. The indirect Coombs test tests for the presence of anti-Rh antibodies in the mother's serum. The figure in the question shows red blood cells with anti-Rh antibodies already attached to them, being mixed with Coombs reagent. This is the test described in the fourth paragraph of the passage as being the direct Coombs tests. Therefore, choices (C) and (D) can be eliminated. The outcome shows that Coombs reagent reacted with the red blood cells; the results of this test are therefore positive, so choice (A) is wrong and choice (B) is right. In order for either a direct or indirect Coombs test to be negative, the anti-Rh antibody cannot be present. For this to have been a positive indirect Coombs test, there would have had to have been an additional, previous step showing the mixing of the mother's serum with washed red blood cells.

36. B Recall that Rh immunity requires prior sensitization, while ABO immunity does not, because of other naturally occurring antigens. Since the woman in the question has never been pregnant, it's highly unlikely that she has ever been exposed to the Rh antigen, which means that she would not produce antibodies against it. So, choices (C) and (D) are wrong. Since her blood type is B, she would not have anti-B antibodies, but would be expected to have anti-A antibodies because of a naturally occurring antigen that resembles the A antigen. So, choice (A) is also incorrect.

37. C The exogenous substance is called RhoGAM, and is derived from Rh immunoglobulin. RhoGAM is anti-Rh antibody. The theory behind it is that if given prior to delivery—the time at which exposure occurs—rhoGAM will attack and coat the fetal Rh⁺ red blood cells. By coating the fetal cells, the Rh antigen is no longer accessible to the mother's system; it is effectively "hidden." The mother therefore does not produce anti-Rh antibody of her own, the baby's coated red blood cells are removed and destroyed by the mother's immune system, and sensitization has been

prevented. So choice (C) is correct. In practice, rhoGAM is usually administered during the postpartum period, specifically within the first 72 hours following birth. Choice (A), Rh antigen, would only enhance the mother's immune response if given prior to delivery. Exposing her to the antigen would lead to the production of anti-Rh antibody, which would increase the risk of erythroblastosis fetalis for future children. The whole idea of the treatment is to prevent the mother from producing her own anti-Rh antibodies, and thereby prevent sensitization, not to enhance it. Choice (B), an immunosuppressive drug, might be an effective way to suppress the immune response, but it would certainly not be safe. An immunosuppressive drug has a broad action, rather than a specific one, and would affect many aspects of the mother's immune system, not just her ability to produce anti-Rh antibodies in response to the Rh antigen on her baby's red blood cells. In fact, immunosuppressive drugs would make the mother more susceptible to infection and would have adverse side effects. Choice (D), iron pills, are often given to pregnant women as an iron supplement to combat the anemia associated with pregnancy. The anemia is due to the excessive circulatory demands exacted by the fetus and the placenta. Iron pills, however, in no way affect the immune system and would therefore not be effective in preventing an immune response of any sort.

Discrete Questions

38. C We're told that a certain chemical inhibits the synthesis of steroids and we're asked to decide which one of the hormones listed in the answer choices would not be affected if a dose of this chemical was administered to a laboratory rat. In other words, we need to know which of these hormones is not a steroid hormone. Steroids are a type of lipid with a carbon skeleton of four fused rings. Different types of steroids differ in the functional group that is attached to the carbon skeleton. Cholesterol is a very important steroid, and is in fact the precursor steroid molecule from which most other steroids, including steroid hormones, are synthesized. For example, the sex hormones—testosterone, estradiol, and progesterone—which are synthesized in the gonads, are steroid hormones modified from cholesterol. This means that we can rule out choice (D). The adrenal cortex is another gland that secretes a family of steroid hormones called the corticosteroids. The two main types of corticosteroids are the mineralocorticoids, such as aldosterone, and the glucocorticoids, such as cortisol. This means that we can also rule out choices (A) and (B), which leaves us with choice (C), epinephrine. The adrenal medulla secretes the hormone epinephrine in response to any type of stress—good or bad. Epinephrine, along with norepinephrine, are compounds called catecholamines, and are synthesized from the amino acid tyrosine by chromaffin cells in the adrenal medulla; hence, epinephrine synthesis would not be affected by this chemical.

39. C Aerobic respiration cannot occur without oxygen, because oxygen is necessary for the final step of aerobic respiration, which is the electron transport chain. Here's a brief summary of glycolysis, the tricarboxylic acid cycle, also known as the Krebs cycle or the citric acid cycle, and the electron transport chain—the three stages of aerobic respiration. Glycolysis is a series of reactions that lead to the oxidative breakdown of glucose into pyruvate. These reactions occur in the cytoplasm and result in the production of two NADH and a net gain of two ATP. Oxygen is not required for glycolysis; therefore, choice (B) is wrong. In the absence of oxygen—that is, under anaerobic conditions—the pyruvate is reduced to lactate; it undergoes this fermentation step so that NAD^+ can be generated. Hence, choice (D) is wrong. In the presence of oxygen—that is, under aerobic conditions—the pyruvate is oxidized to release the considerable energy still stored in its chemical bonds. The pyruvate is transported from the cytoplasm into the mitochondrial matrix. The oxidative decarboxylation of pyruvate into acetyl CoA is catalyzed by pyruvate dehydrogenase complex; one NADH is generated during the formation of acetyl CoA. Next, the acetyl group from the acetyl CoA combines with oxaloacetate, forming citrate. Through a complicated series of reactions, the citrate is completely oxidized, two molecules of carbon dioxide are released, and NADH, $FADH_2$, and ATP are generated. Next, all of the molecules of NADH and $FADH_2$ generated during glycolysis, pyruvate decarboxylation, and the TCA cycle, transfer their high potential electrons to a series of carrier molecules located in the inner mitochondrial membrane. A series of redox reactions is coupled with the phosphorylation of ADP; this is known as oxidative phosphorylation. As the electrons are transferred from carrier to carrier, free energy is released, which is then used to produce ATP. So, choice (A) is also wrong. The final carrier of the electron transport chain, cytochrome a_3, passes its electrons to the final

electron acceptor, molecular oxygen—or O_2. Oxygen picks up a pair of hydrogen ions, forming water. The final ATP tally for aerobic respiration is a net gain of 36 ATP. This is eighteen times as much ATP as is produced during anaerobic respiration. Choice (C) is the only choice which is true and consistent with the observations made by the biochemist in the question.

40. B Acromegaly is a condition that results from an oversecretion of growth hormone. In addition, you're told that growth hormone decreases the sensitivity of insulin receptors. Therefore, to answer this question you've got to determine the physiological effects of an insensitivity to insulin. You should know that insulin is secreted by the pancreas in response to a high blood glucose concentration, and that it lowers blood glucose by stimulating the conversion of glucose into its storage form, glycogen. This means that the cell surface receptors of acromegaly patients do not bind insulin as much as normal cells. Therefore, the effects of insulin will be diminished in an acromegaly patient. If insulin cannot exert its effects on the cells, excess glucose will not be converted into glycogen. Hence, the patient will have excess glucose—in other words, a high blood glucose concentration. Therefore, choice (B) must be the correct answer and choice (A) must be the wrong answer. Choice (C) is wrong because a high blood glucose concentration leads to the excretion of glucose, along with the loss of water in the urine, meaning that patients with acromegaly will have an increased, not a decreased, urine volume. Choice (D) is wrong because cardiac output is defined as the volume of blood pumped by the heart per unit time, which is not even directly related to the sensitivity of insulin receptors.

Passage VI (Questions 41–45)

41. C The genotype of the red male in Cross 6 is either little b, w or little b, little b. Cross 6 is between a red male and a red female, and their offspring are 100% red. What we need to do is look at the offspring and work backwards to determine the genotypes of their parents. As we've just determined, there are two possible genotypes that correspond to a red phenotype—little b, little b and little b, w, since the b allele is dominant to the w allele. With these two genotypes, there are three possible types of crosses (you might want to write these down): little b, little b × little b, little b; little b, little b × little b, w; and little b, w × little b, w. Note that we're not taking into account the gender of the parents: We're not bothering with the fact that the little b, little b × little b, w cross can occur in two different ways—the male can be little b, little b and the female can be little b, w, and vice versa. All of the offspring of Cross 6 are red; therefore, we can eliminate the possibility of a little b, w × little b, w cross since 25% of their offspring would be white. We're left with the other two crosses, both of which produce 100% red offspring. If the male is little b, little b, then the female can be either little b, little b or little b, w; and if the male is little b, w then the female must be little b, little b. Hence the male can either be little b, little b or little b, w and the correct answer is choice (C).

42. A If a large number of the brown offspring from Cross 8 are mated with each other, 6.25% of the offspring are expected to be white. The first thing to do is to figure out the genotypes of the parents in Cross 8. Cross 8 is between a brown male and a red female. The fact that 25% of their offspring are white indicates that both parents are heterozygotes, since white fur is a recessive trait. This means that the genotype of the brown male must be big B, w, and the genotype of the red female must be little b, w. Now we need to figure out the genotypes of the brown offspring. A cross between this brown male and red female results in 25% ww offspring, which are white; 25% little b, w, which are red; 25% big B, w, which are brown; and 25% big B, little b, which are brown. So there are two different brown genotypes in the offspring—big B, little b and big B, w. If a large number of these brown offspring are mated with each other, there are four different crosses possible: (1) big B, little b × big B, little b; (2) big B,w × big B. w; (3) big B, little b × big B, w; and (4) big B, w × big B, little b (this is the second way that the third cross can occur). Now we need to figure out what percentage of the offspring produced in these crosses will have white fur. If we work out the Punnett squares, we find that neither the first, the third, nor the fourth crosses yield any white rats. On the other hand, the second cross, big B, w × big B, w, yields 25% white offspring. But 25% is not your answer. You have to take into account that only one-fourth of the total number of possible crosses between two brown rats yields 25% white offspring. One-fourth of 25% is 6.25%, which is the correct answer, choice (A).

43. D The genotype of the female in Cross 5 can either be big B, big B; big B, little b; or big B, w. In Cross 5, a brown male is mated with a brown female, and we're told that 100% of their offspring are brown. Since we know that the big B allele is dominant to

the little b and w alleles, there are three types of genotypically distinct crosses between two brown parents that result in all brown progeny: (1) big B, big B × big B, big B; (2) big B, big B × big B, w; and (3) big B, big B × big B, little b. From this we see that at least one of the parents must be big B, big B. If the female is big B, big B, then the male can be big B, big B; big B, little b; or big B, w. Likewise, if the male is big B, big B, the female can be big B, big B; big B, little b; or big B, w. So, choice (D) is the correct answer.

44. C The ratio of red to white progeny in a cross between a white male and the heterozygous red female from Cross 9 is 1:1. This question is really quite simple because we're told that the red female has the heterozygous genotype, which we know is little b, w. We don't have to look at the progeny of Cross 9 and try to work backwards from there. Since white fur is recessive, the white male in the cross must have the genotype ww. So the cross is little b, w × w, w. Fifty percent of the offspring have the genotype b, w, which, phenotypically, is red fur; the other 50% of the offspring have the genotype w, w, which corresponds to white fur. Thus, the ratio of red offspring to white offspring is 50% to 50%, which is the same as 1: 1, choice (C).

45. B The ratio of progeny in a cross between a big B, little b male and a pink female would be 2 Brown: 1 Red: 1 Pink. What we need to do is figure out the genotype of the pink female. That's not so hard, since the question stem tells us that rats with pink hair have one copy of the red allele—little b, and one copy of the white allele—w. So, the genotype of the pink female would be little b, w. So if this pink female is crossed with a brown male, the cross is big B, little b × little b, w. If we work out the Punnett square, 25% of the progeny would be big B, little b, which is phenotypically brown; 25% would be big B, w, which is also brown; 25% would be little b, little b, which is red; and finally, 25% would be little b, w, which is pink if these two alleles are incompletely dominant. Thus, the phenotypic ratio of the offspring is 2 Brown: 1 Red: 1 Pink, choice (B).

Discrete Questions

46. A You are presented with an unlabeled figure of a eukaryotic cell and asked to determine which of the four numbered structures would be most affected by a drug that inhibits ribosomal RNA synthesis. So basically, you have to know which structure is involved in the synthesis of ribosomal RNA. The nucleolus is the organelle responsible for ribosomal RNA synthesis, and in the figure, structure 1 is the nucleolus, and so choice (A) is correct. Structure 2 is the nucleus; structure 3 is the Golgi apparatus; and structure 4 is a mitochondrion. Therefore, choices (B), (C), and (D) are wrong, and choice (A) is the right answer.

47. D An exocrine gland is one that excretes its products into tubes or ducts that typically empty onto epithelial tissue, while an endocrine gland is one that releases hormones directly into the bloodstream. The pancreas functions both as an endocrine gland and an exocrine gland. As an endocrine gland, it produces and secretes three hormones: insulin, glucagon, and somatostatin. Insulin lowers blood glucose levels by stimulating the uptake of glucose into tissues, and its subsequent conversion into its storage form, glycogen. So, choice (B) is wrong. Choice (A) is incorrect because it is glucagon that raises blood glucose levels by stimulating the conversion of glycogen into glucose. Somatostatin suppresses both insulin and glucagon secretion. Choice (C) is incorrect because thyroid hormones are involved in the regulation of metabolic rate. As an exocrine gland, the pancreas secretes enzymes that are involved in protein, fat, and carbohydrate digestion; all of its exocrine products are secreted into the small intestine. Pancreatic amylase hydrolyzes starch to maltose; trypsin hydrolyzes peptide bonds and catalyzes that conversion of chymotrypsinogen to chymotrypsin; chymotrypsin and carboxypeptidase also hydrolyze peptide bonds; and finally, lipase hydrolyzes lipids. Thus, choice (D) is the correct answer.

48. C The cerebellum is part of the hindbrain, which is the posterior part of the brain and consists of the pons and the medulla oblongata, in addition to the cerebellum. The cerebellum receives sensory information from the visual and auditory systems, as well as information about the orientation of joints and muscles. In fact, one of the cerebellum's main functions is hand-eye coordination. It also receives information about the motor signals being initiated by the cerebrum. The cerebellum takes all of this information and integrates it to produce balance and unconscious coordinated movement. Damage to the cerebellum could damage any one of these functions. Total destruction of the cerebellum would eliminate all of them. Therefore, choice (C) is correct because destruction of a cat's cerebellum would seriously impair coordinated movement in the cat. Urine formation, choice (A), is the primary function of the kidneys, with a little bit of hormonal regulation to help things out. Choice (B), sense of smell, or olfaction, is a function of the cerebrum not the cerebellum, while thermoregulation, choice (D), is a function of the hypothalamus, which is a part of the cerebrum.

49. D A cell with a high intracellular potassium concentration is placed in a medium with a low potassium concentration and a high ATP concentration. The cell membrane is impermeable to potassium. If the cell were permeable to potassium, you would expect potassium to flow out of the cell, along its concentration gradient. This is clearly not the case: instead, the extracellular concentration of both the potassium and the ATP decreases, while the intracellular concentration of potassium increases. This implies that the potassium moved into the cell, across a membrane that is impermeable to it, and against its concentration gradient. What's the most likely explanation for this phenomenon? Choice (A) says that the potassium passively diffused into the cell, and choice (B) says that the potassium entered the cell by way of facilitated transport. Since both diffusion and facilitated transport occur along a substance's concentration gradient, not against it, both choices (A) and (B) must be incorrect. Furthermore, the potassium could not possibly diffuse across the membrane, because we know that the membrane is impermeable to it. And finally, neither of these choices accounts for the decrease in extracellular ATP concentration. Choice (C) sounds reasonable, except for two factors. First of all, we're told that the membrane is impermeable, yet ATP is not a very lipid-soluble complex with potassium; it enables the potassium to cross the lipid cell membrane and enter the cell. Second, there is no evidence to support the theory that ATP functions as a carrier molecule, shuttling potassium across cell membranes. ATP is the energy currency used by cells. Energy is stored in its high-energy phosphate bonds, and this energy is made available to cells when ATP is hydrolyzed to ADP and AMP. Thus, choice (C) is also incorrect. Choice (D), however, does make sense; it explains both the movement of potassium into the cell against its gradient, as well as the decreased extracellular ATP concentration. Active transport is the movement of a substance against its concentration gradient with the aid of carrier molecules and energy.

Passage VII (Questions 50–55)

50. C Ribosomes are the organelles responsible for protein synthesis from mRNA transcripts. This process is known as translation. Translation is initiated by the binding of an mRNA transcript to the small ribosomal subunit. If mRNA cannot attach to a ribosome, as is the case in multicopy inhibition, translation cannot occur. So, multicopy inhibition represses translation and choice (C) is correct.

Choice (A) is wrong because replication is DNA synthesis. Multicopy inhibition acts on RNA, not DNA, so it can't block replication. Choice (B) is wrong because IN RNA is produced, which implies that transcription—the production of RNA from a DNA template—did occur. So transcription is not repressed by multicopy inhibition. Choice (D), postranscriptional modification, refers to modifications made to an mRNA transcript after transcription has occurred. Post-transcriptional modifications include removing introns, adding a a poly (A) tail or a 5′ cap.

51. A According to the passage, the dam methylation system regulates the rate of transposition and the activation of Tn10 through the methylation of newly synthesized DNA. The initial lack of methylation on the newly synthesized DNA activates the transposase gene and by enhancing the binding of this enzyme to the transposon itself. Therefore, choice (A) is correct, since if a strain of cells lacked the dam methylation system, the strain's new DNA would not be methylated following replication, and the rate of transposition observed by the researcher would increase.

Choice (B) is wrong because multicopy inhibition inhibits Tn10 transposition by ensuring that all IN RNA is bound to OUT RNA and thus cannot be translated. Therefore, if the new strain of cells has an even higher ratio of OUT RNA to IN RNA than is found in normal cells, the chances of IN RNA being translated would be even lower, thereby decreasing the rate of transposition. Choice (C) is wrong because there is no mention in the passage of a relationship between the presence of the duplicated sequence and the rate of transposition. Choice (D) is wrong because increasing the overlap between the IN RNA and OUT RNA would mean that more energy is required to separate the IN RNA from the OUT RNA, which means that the IN RNA would be translated less frequently and the rate of transposition would decrease.

52. B DNA replication is semi-conservative. New DNA consists of an original (parent) strand base-paired to a new strand. Since the original strand is methylated, the DNA is said to be hemi-methylated

(the prefix hemi means half, as in hemisphere). Choice (B) is correct because hemi-methylation allows the cell to distinguish between the two strands and detect (and correct) any mutations or errors that occurred during replication.

Choice (A) is wrong because, although the hemi-methylated DNA arises after DNA replication, it does not affect the rate at which the cell divides. DNA replication, and thus the presence of hemi-methylated DNA, is only an indication that cell division has occurred. Choice (C) is wrong because the use of free methyl groups in the synthesis of amino acids and cofactors is unrelated to the hemi-methylation of DNA. The passage does not discuss either the source of the methyl groups or other possible uses for methyl groups. Choice (D) is wrong because transportation across the nuclear membrane depends on carrier proteins, transport proteins, and the pores found in the nuclear membrane. And although DNA is believed to aid in this transport process, this aid is independent of the methylation.

53. A The basic question posed here is "Can transposition be fatal to the cell?" Given that transposons insert at random sites in the host DNA, insertion within a vital gene is a distinct possibility. Insertion within such a gene would most likely cause a mutation that renders the gene's protein product nonfunctional, which could cause cell death. Thus, choice A is correct.

Choice (B) is wrong because it is OUT RNA that inhibits production of transposon of IN RNA that inhibits production of transposon-specific proteins, not the other way around. Choice (C) is wrong due to faulty logic. The pervasiveness of a transposon does not preclude it being lethal; the two properties are not mutually exclusive. Choice (D) is wrong because the fact that transposon insertion results in the duplication of the target sequence has nothing to do with the viability of the cell.

54. D An understanding of transposons is not required to answer this question. Choice (D) is correct because incubating the cells on agar plates containing ampicillin is the only technique that directly exposes the cells to the antibiotic. It is the only experimental procedure listed that includes what we are trying to measure/determine as a variable. Any cells that contained Tn10 would be able to live in the presence of ampicillin; those that lacked Tn10 would die.

Choice (A) is wrong because, even though a cell with a transposon would be slightly heavier than a cell without one, differential centrifugation is not sensitive enough to detect such a subtle difference in mass and therefore would not be able to distinguish cells on the basis of their ability to resist ampicillin. Choice (B) is wrong because if you were to hybridize the cells with radio-labeled antibody specific for *E. coli* you would only be able to determine which cells were *E. coli*. Choice (C) is wrong because, although staining and microscopy reveal a lot about cell morphology, there is no way to determine if a cell is resistant to certain drugs just by looking at it under a microscope.

55. C RNA is structurally similar to DNA, except that its sugar is ribose and it has uracil (U) instead of thymine (T). So choice (B) and choice (D) can be immediately eliminated, since they include thymine instead of uracil. Cytosine (D) still pairs with guanine (G), while adenine (A) pairs with U instead of T. According to the passage, OUT RNA pairs with IN RNA, and since RNA (like DNA) has a 5' end and a 3' end, when the two RNA molecules base pair, they align in an anti-parallel orientation—the 5' end of one molecule pairs with the 3' end of the other molecule, and vice versa. Therefore, a strand of RNA with the sequence 5'-AUAUGCC-3' will base pair with a strand of RNA with the sequence 5'-GGCAUAU-3'. So choice (C) is correct. Choice (A) is wrong because although the sequence is correct, the polarity is wrong. Note: It is convention to list DNA and RNA sequences in the 5'–3' direction.

Passage VIII (Questions 56–60)

56. C The initial stage of protein synthesis is transcription, which is the synthesis of mRNA using DNA as a template. So it seems odd that the half-egg without the nucleus was able to synthesize any protein at all. However, if the whole egg that would eventually be split into the two half-eggs had transcribed the same amount of mRNA during maturation as the other whole egg, then protein synthesis would be possible in all three eggs after activation, since translation occurs in the cytoplasm. The amount of protein initially synthesized in the two half eggs would be comparable to the amount synthesized by the whole egg. So, choice (C) is correct. Choice (A) is wrong because the half-egg without the nucleus can't synthesize mRNA after activation, since transcription occurs in the nucleus. Choice (B) is wrong because, even if all the eggs did synthesize

and amass the same amount of amino acids during maturation (when all of the eggs had nuclei), this could not account for the experimental results. A pool of amino acids will not spontaneously form proteins in the absence of transcription and translation. Transcription must have occurred prior to activation. Choice (D) is wrong because the half-egg without the nucleus would not be able to synthesize any rRNA after activation, since rRNA is synthesized in the nucleolus, which is found within the nucleus.

57. A Protein synthesis in all of the eggs was approximately the same for the period of time shown in Figure 7 because the mRNA needed for early development was transcribed during maturation and stored in the cytoplasm in an inactive form until the eggs were activated. (See the explanation to the previous question). Following activation, these mRNA transcripts were translated into the proteins necessary for the initial stages of embryonic development. However, active mRNA transcripts have a very short life-span—more mRNA must be transcribed for protein synthesis to continue. But the half-egg without the nucleus cannot do any further transcribing, since nuclear DNA is a prerequisite. So if the experiment had run any longer, the level of protein synthesis in the half-egg without the nucleus would have dropped dramatically after the initial mRNA transcripts had been translated. Since the whole egg and the half-egg with the nucleus are capable of continuing transcription, protein synthesis would continue to increase in similar patterns. Choice (A) is correct because it is the graph that accurately depicts these results.

58. B The phrasing of the question stem implies that the binding of sperm to receptors found on the outer surface of the egg is necessary for fertilization. Choice (B) is therefore correct, since a drug that blocked sperm access by irreversibly binding to these receptors would prevent the sperm from penetrating the egg, and thus prevent fertilization. A drug or device that blocks or inhibits fertilization is called a contraceptive.

Choice (A) is wrong because, while it is true that if an egg is not fertilized it will eventually atrophy, this is not a direct consequence of blocked surface receptors. An ovulated, unfertilized egg has a specific life-span, independent of the accessibility of its surface receptors. Choice (C) and choice (D) are wrong because they state that a drug that irreversibly blocked these receptors would not be considered an

KAPLAN
Test Prep and Admissions

effective contraceptive. Choice (C) can be ruled out because artificial activation is unrelated to contraception. Choice (D) can also be ruled out because the sperm nucleus can't fuse with the egg nucleus until the sperm has penetrated the outer layer.

59. B The three egg types (gametes) used in the experiment are all haploid cells. This means that each of the eggs has only one of each pair of chromosomes in its nucleus—this is the result of meiosis: meiosis halves chromosome number, fertilization restores chromosome number, and mitosis maintains it. So the whole egg is a haploid cell, and therefore contains half the number of chromosomes as an autosomal cell. An autosomal, or diploid, cell is a cell that has two copies of every chromosome—one of maternal origin, one of paternal origin. For example, in humans, a diploid cell such as a liver or pancreas cell, contains 22 pairs of autosomes and a pair of sex chromosomes, for a total of 46 chromosomes. Likewise, the half-egg with a nucleus is also a haploid cell, since its nucleus is the product of meiosis. So, the half-egg with a nucleus has the same number of chromosomes as the whole egg, and half the number of chromosomes as an autosomal cell. The half-egg without a nucleus has no chromosomes. Thus, choice (B) is correct and choice (A) is wrong. Choice (C) and choice (D) are wrong because using terms "half" and "twice" are meaningless when one is comparing chromosome number between the half-egg with a nucleus and the half-egg without a nucleus, since the latter has zero chromosomes.

60. D Mammalian embryos, which develop internally, obtain the majority of their nutritional requirements directly from their mother's blood via the placenta. Yolk plays a minor role in terms of nourishment for these embryos. With externally developing embryos, however, the mother cannot provide direct nourishment. The only source of nourishment for an externally developing animal is the egg yolk. Yolk is rich in protein and serves as a store of nutritional reserves for the egg. Therefore, it makes sense that yolk should occupy a much greater percentage of total egg volume in an externally developing egg than in an internally developing egg. So, choice (D) is correct.

Choice (A) is wrong because, although eggs that develop externally are susceptible to predation, it is typically factors such as camouflage and shell strength that protect the developing embryo, not egg size. Furthermore, the question stem does not address the issue of a comparison of actual egg size between externally developing and internally developing eggs. Choice (B) is wrong because animal size is independent of the percentage of total egg volume occupied by yolk whether the animal developed externally or internally. Choice (C) is wrong because it wouldn't account for the difference in yolk volume. Furthermore, gestation is defined as the duration of internal development, which is unique to mammals and marsupials. And typically, the period of time spent in the womb is generally much longer than the period of time it takes for an egg to hatch: human gestation is 9 months, giraffe gestation is 14 months, and elephant gestation is 21 months, while it takes only 21 days for a chick to hatch.

Discrete Questions

61. B The nucleus of a zygote contains all of the genetic information needed by all of the future cells of the adult organism. As the zygote divides and cells differentiate, each cell inherits the same exact DNA. The difference lies in which genes are expressed and which are repressed. A tadpole intestinal cell is a type of differentiated cell. According to the question stem, when the nucleus of a differentiated intestinal cell is transplanted into a denucleated from zygote, the frog develops normally. Choice (B) is therefore correct. If the salvaged nucleus had lacked genetic material essential to a developing tadpole or had been unable to repress the genes that, when expressed, lead to the formation of an intestinal cell, the zygote would have at best developed into a mass of intestinal cells.

Choice (A) is wrong because the experimental results prove that irreversible gene expression does not occur, otherwise the transplanted nucleus would not have been able to direct the development of the zygote into a normal frog. Choice (C) is wrong because the results of the experiment neither support nor contradict this claim. DNA is the known carrier of genetic information. Choice (D) is wrong because, as with choice (C), it is not addressed by the experiment. Furthermore, all eukaryotic ribosomes are the same and are found only in the cytoplasm of a cell, not the nucleus.

62. D Osmotic pressure is the tendency of water to diffuse from an area of lower solute concentration to an area of higher solute concentration. Albumin is an osmotically active plasma protein and is one of the primary determinants of the blood's osmotic (or oncotic) pressure. If a membrane is impermeable to a particular solute, then water will diffuse across the membrane until the differences in the solute concentrations have been equilibrated. An infusion of albumin would create a greater concentration of albumin in the capillaries than in the interstitial fluid that bathes the capillaries. Water would therefore move from the interstitial fluid into the capillaries. Therefore, choice (D) is correct, because there would be a decrease in the movement of water from the capillaries to the interstitial fluid.

Choice (C) is wrong, because there would be a decrease, not an increase, in the movement of water from the capillaries to the interstitial fluid. Choice (A) and choice (B) are wrong, since increasing or decreasing a solute's plasma concentration via infusion does not normally affect a cell's permeability to that solute.

63. A A young person complaining of thirst, polyuria, and weight loss is the classic presentation of Type 1, or juvenile, diabetes mellitus. Juvenile diabetes occurs when, for an unknown reason, the β cells of the pancreas stop producing enough insulin. Insulin, which is secreted by the pancreas, lowers blood glucose by promoting glucose uptake and the conversion of glucose to glycogen. Insulin is the only glucose-lowering hormone. Choice (A) is correct, because if insulin was low, blood glucose would be elevated.

ACTH (adrenocorticotropic hormone) is the anterior pituitary hormone that stimulates the adrenal cortex to secrete glucocorticoids. Glucocorticoids raise blood glucose levels and decrease protein synthesis. Choice (B) is wrong, since low ACTH would decrease glucocorticoid secretion, thereby lowering blood glucose. Glucagon, another pancreatic hormone, is antagonistic to insulin's effects: glucagon stimulates the conversion of glycogen to glucose in the liver, thereby raising blood glucose. So, choice (C) is wrong, since blood glucose would be low if glucagon levels were low. Choice (D) is wrong, because cortisol, a steroid hormone, has glucocorticoid activity, and as previously mentioned, low glucocorticoid levels will result in low blood glucose.

Passage IX (Questions 64–67)

64. D The sequential model assumes that there is not an equilibrium between R and T and that the

conformational state of one subunit can be different than the conformational state of the protein's other subunits. This means that intermediate forms of a protein can exist: one subunit can be in the R form while others remain in the T form. In the concerted model, an equilibrium between the two conformational states is assumed and all of the subunits of a protein must be in the same conformational state at any given time. This means that if one subunit of a protein is in the R form, then all of the subunits in that protein molecule must be in the R form.

Since we are dealing with a protein that consists of two identical subunits, the possible configurations are TT, RR, and RT. RT is the only one that could be found in the sequential model but not the concerted model. Therefore, choice (D) is correct. Choice (A) and choice (B) are wrong because these configurations are consistent with both models. Choice (C) is wrong, since this would be a configuration for a protein consisting of four subunits, not two.

65. B When a protein obeys the concerted model, if one of the subunits binds substrate—thereby converting it to the R form—all of the other subunits in the protein also convert to the R form. In other words, there will be subunits in the R form without bound substrate. So at nonsaturating succinate concentrations, there will be a greater fraction of ATCase subunits in the R form—with and without succinate bound—than the fraction of ATCase with succinate bound. This means that the f_R curve should be to the left of the Y curve. As you approach saturating concentrations of succinate, the rest of the active sites fill with substrate, and therefore the f_R and Y curves should begin to converge. When all of the active sites are 100% filled, the two curves should meet, because at this point, every subunit in the R conformation has substrate bound. Therefore, choice (B) is the correct answer.

Choice (A) is wrong because cooperative binding is consistent with both the concerted model and the sequential model. Choice (C) and choice (D) are wrong because if ATCase conformed to the sequential model, every active site occupied by succinate would be in the R form, while the other subunits would remain in the T form, until their active sites bound succinate. In other words, the fraction of active sites in the R form would equal the fraction of active sites with substrate bound, and the two curves -f_R and Y—would be identical.

66. C The sequential model allows for intermediate forms while the concerted model does not; only the concerted model assumes conformational equilibrium between the two forms; and only in the sequential model can the binding of substrate to one subunit actually decrease the substrate-binding affinity of the other subunits in a molecule. Choice (C) is therefore correct, since it supports the sequential model only.

Choice (A) is wrong because it supports the concerted model only. Choice (B) is wrong because it supports both models, since the addition of a saturating concentration of substrate would mean that all subunits would have substrate bound, and therefore all subunits would be in the R form. Choice (D) is wrong because it doesn't really support either model. Besides, denaturation would prevent any protein from binding substrate.

67. B Since hemoglobin conforms to the concerted model, it exists in either the T form or the R form; intermediate forms are not possible. This means that after the first subunit of hemoglobin molecule binds oxygen and converts to the R form, the other three subunits also convert to the R form. The T form has a lower affinity for oxygen than does the R form, and in the presence of high oxygen concentration, the conformational equilibrium will shift toward the R form. So the higher the concentration of oxygen, the greater the ratio of the R form to the T form. Of the four choices, the left ventricle has the highest oxygen concentration and so choice (B) is correct.

Remember, deoxygenated blood is carried to the right atrium by the superior and inferior venae cavae and the coronary veins, which is why choice (D) is wrong. The right atrium pumps this blood into the right ventricle, which is why choice (A) is wrong, and the right ventricle pumps it to the lungs via the pulmonary arteries, which is why choice (C) is wrong. Gas exchange occurs in the lungs, and the oxygenated blood is carried to the left atrium by the pulmonary veins. The left atrium pumps this blood into the left ventricle, which it pumps into systemic circulation via the aorta, and cardiac muscle via the coronary arteries, which are the first branches off of the aorta.

Passage X (Questions 68–72)

68. B Uric acid is formed from the breakdown of purines, which means that hyperuricemia is caused by an increase in purine degradation. Adenine and guanine, along with cytosine, thymine, and uracil, are the nitrogen bases found in nucleic acid. Adenine and guanine are the purine bases, and cytosine, uracil, and thymine are the pyrimidine bases. (The mnemonic "CUT the PY" is a good way to remember which are which: C, U, and T are the PYrimidine bases.) The right answer will be the process that yields purines as its end-product. Choice (B) is correct because suppression of nucleotide degradation would decrease purine recycling, which would decrease uric acid production, which would alleviate hyperuricemia. Nucleotides, the building blocks of nucleic acid, consist of a nitrogen base, a sugar, and a phosphate group, and so nucleotide degradation would yield purines (as well as pyrimidines).

Choice (A) is wrong because fatty acids are long chain carboxylic acids. Fatty acid catabolism would therefore not yield purines. Choice (C) is wrong because amino acid synthesis would obviously not yield purines. Choice (D) is strong since cholesterol synthesis produces cholesterol.

69. C A substrate analog, such as allopurinol, is a substance with a structure similar to the natural substrate of an enzyme and which, because of this similarity, inhibits the action of the enzyme. Many competitive inhibitors are substrate analogs. A competitive inhibitor competes with an exnzyme's substrate for the enzyme's active site. If the enzyme binds the analog, enzyme function is effectively inhibited. In terms of this question, the enzyme is xanthine oxidase and its natural substrate is hypoxanthine. As an analog of hypoxanthine, allopurinol acts as a competitive inhibitor, binding to xanthine oxidase's active site, and inhibiting the conversion of adenine into urate. Thus, choice (C) is correct.

Choice (A) is wrong because both allopurinol and hypoxanthine are the substrates—but the substrates don't bind to each other, they bind to the enzyme active sites. Although it is true that allopurinol binds to xanthine oxidase, this binding inhibits rather than promotes the conversion of hypoxanthine into xanthine; so choice (B) is wrong. Furthermore, the conversion of adenine to hypoxanthine is catalyzed by another enzyme, one that is not identified in the passage or figure. The binding of allopurinol to xanthine oxidase does not block this reaction, and so choice (D) is wrong.

70. A Since the allele for HGPRT deficiency is an X-linked recessive, a man with HGPRT deficiency would have the genotype XhY, while a woman with the deficiency would have the genotype XhXh. If the woman was a carrier of the disease, but did not have the disease herself, her genotype would be XhX. So in a cross between a normal man (XY) and an HGPRT-deficient woman, (XhXh), the possible F1 genotypes are: 50% XhX and 50% XhY. All daughters will be carriers (none will inherit the disease itself) and all sons will inherit the deficiency. Therefore, choice (A) is the correct answer since the question asks only about female children.

71. A The key word in the question stem is migration. Migration implies movement, and cell movement is mediated by microfilaments and microtubules. Microfilaments are microscopic filaments composed chiefly of actin. Microfilaments function in cell support and in movement. Microtubules are microscopic tubular structures composed chiefly of tubulin and serve similar functions. In addition to being found in the cytoskeleton, microtubules are fibers of mitosis. Choice (A) is correct, because colchicine disrupts mitosis by inhibiting microtubule polymerization. The inhibition of granulocyte migration reduces inflammation of gout because granulocytes are the type of white blood cells responsible for the inflammatory response.

Choice (B) is wrong, because ribosomes are the organelles responsible for translating mRNA transcripts into peptide chains. Choice (C) is wrong, because the nucleolus is the suborganelle of the nucleus responsible for the synthesis of rRNA. Choice (D) is wrong, because the Golgi apparatus is the organelle responsible for the modification and packaging of proteins for export.

72. D Hypouricemia means having an abnormally low concentration or urate. According to the passage, urate lowers the rate of mutation by decreasing the rate of attack on DNA by highly reactive species such as free radicals and superoxide anions. Since DNA mutations are the fundamental event in carcinogenesis, mutations dramatically increase the risk of cancer. Therefore, if a patient has a low urate level, their rate of mutation will be increased, and hence the risk of cancer will also increase. Therefore, choice (D) is correct.

Choice (A) is wrong because there's no mention in the passage of increased risk for bacterial infection as being a direct consequence of hyper- or hypouricemia. Choice (B) and choice (C) are wrong, since according

to the passage, elevated urate (hyperuricemia) results in gout, which is often characterized by arthritis. A person with hypouricemia would not suffer from gout or develop any of its related symptoms.

Discrete Questions

73. A According to the law of independent assortment, which holds only for unlinked genes, the probability of a particular cross producing an individual, with a particular genotype is equal to the product of the probabilities of inheriting the individual genotype specified for each allelic pair. For example, there are two individuals that are heterozygous at the loci for genes A, B, and C (genotype = AaBbCc) and you want to know the probability that a cross between these two will produce an offspring that is homozygous dominant at the three loci (genotype = AABBCC). Both parents are heterozygous at the A locus (As), which means that the ratio of their offspring with respect to the A locus is 1AA : 2Aa : 1aa. Since both parents are also heterozygous at the B and C loci, the probability of their offspring being BB is 1/4, and the probability of their offspring being CC is also 1/4. Therefore, the probability that these two individuals will have an offspring with the genotype AABBDCC is equal to the product of the individual probabilities $1/4 \times 1/4 \times 1/4 = 1/64$; so choice (A) is the correct answer.

74. C Cyanide is a poison that interferes with the electron transport chain of the inner mitochondrial membrane by binding to one of the electron transfer complexes. In doing so, cyanide completely inhibits the flow of electrons and effectively paralyzes the transport chain. As a consequence, the proton pump comes to a stop, and ATP cannot be generated aerobically. In addition, electron carriers such as NADH and FADH2 can't deliver their high energy electrons to the electron transport chain because it has become backed up. This means that NAD^+ and FAD are not regenerated, and aerobic respiration cannot continue. Energy is required for a phage, or any organism for that matter, to replicate. DNA replication, RNA transcription, and protein synthesis all require ATP and are processes necessary for viral replication, as well as for host cell function. If aerobic ATP formation is inhibited, there will not be enough ATP for normal cell function or for viral replication. Therefore choice (C) is correct. Choice (A), choice (B), and choice (D) are wrong because they do not correctly describe cyanide's mechanism of action.

GRE Biology Practice Set

1. Plasmids (circular genetic elements present in bacteria) can:

   A. act like transposons.
   B. be found in all eukaryotic cells.
   C. replicate only when other plasmids replicate.
   D. never move from cell to cell.
   E. only be made from RNA.

2. It has been observed that neighboring sections of a polypeptide chain can take shape in such a way as to block the access of water to certain areas of the polypeptide surface. This indicates that:

   A. water bonds to protein surfaces using hydrogen bonding.
   B. different conformations of the same protein may alter the protein's specificity or reactivity.
   C. enzyme active sites do not work properly when exposed to water.
   D. enzyme-substrate interactions depend upon exposure to water only on select areas of the tertiary structure.
   E. the interior of the protein is entirely hydrophobic.

3. It is important that certain free ribosomes bind to the outer surface of the endoplasmic reticulum (ER) in order to complete their protein synthesis because:

   A. the ER membrane will break down without the presence of numerous ribosomes.
   B. it allows for the synthesis of certain proteins to be completed in the cytosol.
   C. it prevents the possibility that the synthesis of certain proteins, such as lysosomal hydrolases, would go to completion in the cytoplasm.
   D. mitochondrial ribosomes must transcribe proteins encoded for by mitochondrial DNA in this manner.
   E. posttranscriptional modifications to the mRNA, such as the addition of a poly-A tail, could not take place outside of the ER lumen.

4. Heat-stable DNA polymerases that have $3' \rightarrow 5'$ exonuclease activity would be most useful during polymerase chain reactions (PCR) for which of the following reasons?

   A. The enzymes would stabilize the hydrogen bonds between the bases of each strand.
   B. The enzymes do not break down during the gel electrophoresis process that typically takes place after PCR in order to isolate DNA segments of interest.
   C. These polymerases replicate DNA much faster than normal polymerases that do not have exonuclease activity.
   D. The amplification of a PCR product depends upon the attachment of PCR primers, which anneal to complementary sequences on the DNA template.
   E. PCR uses heat to denature DNA and the rapid rate of DNA replication often results in copying errors.

5. Although all B-lymphocytes start out with antibody molecules bound to the outer surface of their cell membranes, an immune response often results in the secretion of free antibodies by B-cells rather than the attachment of antibodies into the plasma membrane. The mechanism causing this change most likely involves:

   A. a change in the types of lipids used to build new plasma membrane after B-cell activation.
   B. alteration of the heavy and light polypeptide chains used to build the antibodies.
   C. changes in RNA processing resulting in the addition or removal of certain signal sequences on the RNA coding for antibody proteins.
   D. mutations in the V, D, and J regions of the chromosomes that code for antibody protein structure.
   E. binding of B-cells to helper T-cells within lymph nodes.

6. The membrane-spanning regions of transmembrane proteins are frequently α-helical. This is best explained by the fact that:

   A. the α-helix is the native conformation with the greatest stability out of all possible 3-D configurations of transmembrane proteins.
   B. β-pleated sheets are alternative conformations only when the membrane possesses many cholesterol molecules for added reinforcement.
   C. the polarity of the space between the lipid bilayers requires maximal exposure to charged amino acid side chains within the protein's structure.
   D. the α-helix shields polar groups on amino acid side chains within the core of the transmembrane region.
   E. α-helices resemble phospholipids with their alternating polar and nonpolar regions.

7. Many neighboring animal cells have connections between them that serve as direct passageways between their cytoplasms, allowing the movement of ions and small molecules back and forth. These connections also couple electrical responses in one cell with electrical responses of adjacent cells. These cell-to-cell linkages are known as:

   A. plasmodesmata.
   B. gap junctions.
   C. hemidesmosomes.
   D. lamella.
   E. occluding junctions.

8. The antigen-specific cytotoxic T cell of the vertebrate immune system represents a main line of defense against invading pathogens. All of the following are characteristics of the cytotoxic T cell EXCEPT:

   A. variable, antigen-specific receptors lie on the outer surface of its cell membrane.
   B. it is able to recognize foreign peptides complexed to MHC glycoproteins on the surfaces of other cells.
   C. proteins capable of lysing bacterial cell membranes are released by it in order to kill infected cells.
   D. it forms identical T cell clones after immune system stimulation.
   E. the ability to complex directly with viral particles and bacteria before releasing cytotoxic compounds results in a high degree of accuracy in targeting.

9. Hox genes are homeotic genes that encode differences in the development of various organs and tissues of an embryo. The Hox genes of chordates, arthropods, and nematodes have been sequenced and studied to reveal a high level of similarity across these animal phyla. Most chordates, however, have four *"trans-Abd-B"* genes, known as Hox 10-13, that arthropods and nematodes do not have. These genes are expressed in the:

   A. anterior nerve ganglia that develop into the brain.
   B. muscular post-anal tail that extends posteriorly.
   C. tissues that develop into the ventral nerve cord.
   D. muscle cells of the heart and blood vessels.
   E. tissues that are segmented by joints.

10. The resting membrane potential of neurons depends upon all of the following EXCEPT:

    A. differential distribution of ions across the axon membrane.
    B. the selective permeability of the axon membrane.
    C. saltatory conduction via nodes of Ranvier.
    D. negatively charged phosphate groups on nucleic acids.
    E. Na^+/K^+ ATPase transmembrane pumps.

11. Which of the following features is found in annelids but not in echinoderms?

    A. A true coelom
    B. Water vascular system
    C. Specialized cells and tissues
    D. Gills
    E. Segmentation

12. Exposing certain types of cells to the hormone prolactin results in:

    A. the secretion of carbohydrate-digesting disaccharidases from the pancreas.
    B. the stimulation of milk production.
    C. a decrease in follicle maturation in females.
    D. an increase in blood sugar concentration.
    E. the stimulation of bone and muscle differentiation and growth.

13. The plant hormone responsible for inhibiting growth and for closing the stomata during periods of water stress is:

    A. auxin.
    B. gibberellin.
    C. abscisic acid.
    D. choline.
    E. cytokinin.

14. In some populations of organisms, the number of adults in the population may be greater than the number of adults actually contributing genes to the next generation. This situation would:

    A. increase the calculated genetic drift of the population because the size of the population is effectively smaller than it seems.
    B. increase the number of heterozygotes in the population and, thus, the amount of variation within the population.
    C. contribute to disruptive selection within the population and the formation of two separate subspecies.
    D. decrease the probability that the frequency of certain alleles possessed by nonmating members of the population would drift toward zero.
    E. result from monogamous relationships between males and females of the population.

15. Among certain populations of salamander in California exist subspecies that have overlapping geographic ranges and frequently encounter each other, although mating between the populations does not always occur. These subspecies would be considered:

    A. allopatric.
    B. sympatric.
    C. parapatric.
    D. hybrids.
    E. sexually isolated.

16. When organisms that are homozygous for a particular beneficial allele gradually replace those who possess more harmful alleles within a population, this is known as:

    A. disruptive selection.
    B. Hardy-Weinberg equilibrium.
    C. stabilizing selection.
    D. unstable equilibrium.
    E. directional selection.

17. Inbreeding can sometimes bring an inbred population to a genetic equilibrium in which its fitness is increased over the fitness that existed before inbreeding began. A reasonable explanation for this is that:

    A. inbreeding causes an initial decline in fitness due to the loss of heterozygosity in the affected population which then rises again as heterozygotes increase in number.
    B. self-fertilization increases the mean fitness of many organisms because it results in the loss of variety among the phenotypic features of members of the population.
    C. new allele frequencies after the population stabilizes will reflect only successful genotypes.
    D. deleterious alleles masked by heterozygosity in earlier populations will be exposed and eliminated due to inbreeding.
    E. most homozygous inbred populations are less fit than other, more variable populations.

18. Both insects and birds use wings to fly, yet at best they share a distant evolutionary ancestor. In reality, the evolutionary lineages of insects and birds diverged perhaps as long as 400,000,000 years ago. The feature of wings that both birds and insects share is considered:

    A. homologous.
    B. analogous.
    C. divergent.
    D. a mutation.
    E. polygnous.

19. A biome that experiences all four seasons—winter, spring, summer, and fall—and is characterized by warm, moist springs and summers is the:

    A. tropical deciduous forest.
    B. taiga.
    C. temperate deciduous forest.
    D. chapparal.
    E. tundra.

20. The move onto land by animals and their success in a variety of biomes could be accomplished only in stages as certain novel traits arose. These traits necessary for the proliferation of life on land included all of the following EXCEPT:

    A. muscular limbs to resist gravity.
    B. the amniote egg for embryonic development.
    C. homeothermy for effective temperature regulation.
    D. moist skin for diffusion and gas exchange.
    E. fur for added warmth and protection.

Question 21

Consider the following blood group data taken from a population in Hardy-Weinberg equilibrium with respect to the alleles responsible for different blood factors. All individuals in the population possess two different blood factors, each coded for by a dominant allele and a recessive allele. For the first blood factor, allele R is dominant to allele r, so both RR and Rr individuals test as blood type R, while rr individuals test as blood type r. For the second blood factor, allele F is dominant to allele f, so both FF and Ff individuals test as blood type F, while ff individuals test as blood type f. The frequencies observed for blood type in the population are as follows:

Table 1

Type	Frequency
RF	0.60
Rf	0.15
rF	0.24
rf	0.01

21. The frequency of the r allele in this population is _____, while the frequency of the F allele in the population is _____.

    A. 0.60, 0.50
    B. 0.25, 0.25
    C. 0.50, 0.60
    D. 0.25, 0.84
    E. 0.50, 0.40

22. Parthenogenesis differs from hermaphroditism as a means of reproduction because:

    A. parthenogenesis is a form of sexual reproduction while hermaphroditism is a form of asexual reproduction.
    B. parthenogenesis usually involves the production of haploid offspring from unfertilized eggs while hermaphroditism produces diploid offspring from a sperm and egg union.
    C. hermaphroditism cannot occur without meiosis, while parthenogenesis can.
    D. hermaphroditism is advantageous in more stable environments where mates can be easily found, while parthenogenesis is advantageous in environments where mates are not easily found.
    E. homeotic genes are not involved in parthenogenetic development, yet play a crucial role in the development of hermaphroditic young.

GRE Biology Practice Set: Answers and Explanations

ANSWER KEY

1. A (See Concept 18.3)
2. B (See Concept 5.4)
3. C (See Concept 17.4, pp. 324–326)
4. E (See Concept 20.1, pp. 391–392)
5. C (See Concept 43.2)
6. D (See Concept 5.4; Concept 7.1, pp. 127–129)
7. B (See Concept 6.7)
8. E (See Concepts 43.2 and 43.3)
9. B (See Concept 24.3, pp. 485–486)
10. C (See Concept 48.2)
11. E (See Concepts 33.5 and 33.8)
12. B (See Concept 45.3, p. 952)
13. C (See Concept 39.2)
14. A (See Concept 23.3)
15. B (See Concept 24.2, pp. 478–480)
16. E (See Concept 23.4, pp. 464–466)
17. D (See Concept 23.3; Concept 52.5)
18. B (See Concept 25.1)
19. C (See Concept 50.4, pp. 1100–1103)
20. D (See Concepts 34.5, 34.6, and 34.7)
21. C (See Concept 23.1, pp. 456–458)
22. B (See Concept 46.1)

EXPLANATIONS

1. A Plasmids are able to reintegrate themselves back into the main bacterial chromosome, much as transposable elements in eukaryotic genomes can move from chromosome to chromosome using enzymes and insertion sequences. Plasmids are not found in eukaryotic cells, choice (B), nor are they made only out of RNA, choice (E). Plasmids can replicate themselves independently of the main bacterial chromosome and they can easily be transferred from cell to cell, so choices (C) and (D) are also wrong.

2. B The fact that certain substances can gain access to certain areas of a protein based upon that protein's 3-D conformation suggests that different conformations can alter protein specificity. While water bonds to many surfaces by hydrogen bonding, choice (A), this is not an adequate explanation for the significance of the water-blocking conformational change described in the question stem; nor is

there any evidence offered to support choices (C) or (D). In choice (E), it is possible for the interior of the protein to be entirely hydrophobic, but again there is no evidence for this in the question stem. Beware of extreme-sounding answers such as choice (E).

3. C Many proteins must be deposited into the ER lumen (membranous sacs) as they are made. Some of these are to be secreted out of the cell and must start their journey in the ER; others are simply too dangerous to synthesize in the cell's cytoplasm (cytosol), such as lysosomal hydrolases that would digest away parts of the cell if allowed to freely float around the cell after synthesis. Thus, choice (C) is the only correct answer here.

4. E Although PCR needs heat to break apart DNA for replication, the heat can often inactivate the enzymes needed for the replication. Therefore, heat-stable DNA polymerases are extremely beneficial, especially when they possess an exonuclease proofreading ability that cuts down tremendously on errors made during copying.

5. C The most sensible answer here is that changes in RNA processing are what cause the release of free antibody proteins rather than their attachment into the cell membrane of the B-cells. Differences in signal patches or signal peptides would allow for changes in the targeting of these antibodies.

6. D The α-helical structure of membrane-spanning regions is useful to shield the polar amino acid side chains from the hydrophobic interior of the cell membrane. Pleated sheets or other conformations would simply not be as stable as they passed through the membrane. However, there is no evidence presented that the α-helix is always the most stable structure, so eliminate (A) for being too extreme an answer choice; nor is there any suggestion that β-pleated sheets have anything to do with cholesterol, so (B) is out as well. There is no polarity of the space between the lipid bilayers, (C), because that space is hydrophobic and nonpolar, and α-helices do not resemble the structure of phospholipids, choice (E)—also, phospholipids do not have alternating polar and nonpolar regions as suggested.

7. B Gap junctions are connections between cells that enable ions and other small material to move between adjacent cells. They are essential for rapid electrical conduction across large tissues or organs, such as in the heart. Plasmodesmata, choice (A), occur only in plants, and lamella, choice (D), are inner sections of plant cell wall. Try to eliminate answer choices quickly if they are not relevant to the information presented in the question stem—here, the stem mentions "animal cells."

8. E T cells, unlike B cells, cannot complex directly with antigens. They are able to bind to the body's own infected cells by recognizing a combination of foreign antigens displayed on the cell surface AND a self-identity protein called MHC (major histocompatibility complex). All other answer choices are characteristics of T cells.

9. B The one trait of all chordates not found in other phyla, including in Arthropoda and Annelida, is the presence of a muscular, post-anal tail (which, in humans, is merely the vestigial coccyx—a series of fused vertebrae known as the "tailbone.") Arthropods, annelids, and chordates all have bunches of nerve cells (ganglia) at their head ends (anterior ends) that develop into the brain, so (A) is wrong, and chordates do not have ventral nerve cords at all (choice C). All phyla described in the question stem have muscular hearts, even the arthropods that have an open circulatory system; thus, (D) can be eliminated. One characteristic of all arthropods is the possession of jointed appendages, such as limbs. The "trans-Abd-B" Hox genes must, therefore, code for the development of a tail.

10. C Saltatory conduction describes the ability of the nerve signal to "jump" down the length of the axon from unmyelinated section to unmyelinated section. Nodes of Ranvier are spaces between myelin coverings around the axon, and the nodes allow the movement of ions such as Na^+ or K^+ across the nerve cell membrane, whereas the myelinated areas do not. Yet nodes of Ranvier are involved in nerve cell action potentials, not in resting potentials. Choice (C) is correct here.

11. E The GRE loves to ask questions like this, so learn the basic differences between animal taxa (groups). The annelids are also known as the "segmented worms." Both echinoderms and annelids are coelomates, yet only echinoderms possess a water vascular system. Both have specialized cells and tissues and neither has gills. Thus, choice (E) is right.

12. B Even if you had never seen the name of this hormone, you should be able to guess based on the letters "lact" within the name. Think "lactose sugar" and you'll be all set. Lactose is the sugar present in milk and prolactin is a hormone that acts on the mammary glands to stimulate milk production.

13. C This is a factual recall question that tests your ability to recognize the names of the plant hormone responsible for growth inhibition. Abscisic acid is the chemical that causes plants to stop growing and become dormant, particularly in times of stress (such as during a drought). Auxins and gibberellins are growth-promoting hormones, and cytokinins retard aging and control apical growth. Choline is not a plant hormone at all, but rather a building block of the neurotransmitter acetylcholine.

14. A In some populations, one male may mate with many females, especially when members of the population live as harems made of a dominant male and several females. Because only the dominant males end up passing on their genes to the next generation, the population size is effectively smaller than it would be if all males contributed their genes to the gene pool. Thus, genetic drift increases as it typically does with small populations. Genetic drift tends to decrease the number of heterozygotes in a population because it reduces the genetic variation within a population, so choice (B) is incorrect. Because genetic drift weeds out heterozygotes, it allows the population to drift toward the more successful homozygous dominant individuals (called directional selection), so (C) is also wrong. And (D) is out because many alleles possessed primarily by nonmating members will drift toward a frequency of zero within the population over several generations, since these members are not passing on these alleles. Monogamous relationships in (E) do not usually result in a situation where only a few males pass on the genes for the next generation.

15. B Sympatric species are those species whose geographic ranges overlap and who frequently encounter each other. Think of *sym-* meaning "same" and *-patric* meaning "land"—these species occupy the same land. Allopatric species, choice (A), live in distant habitats and will not meet, whereas parapatric species, choice (C), overlap geographic ranges only at their borders, where hybrid species can sometimes be formed.

16. E Movement in one phenotypic direction over time is defined as directional selection, where one extreme phenotype is favored over all other possible phenotypes given a particular environment. Disruptive selection, choice (A), weeds out intermediate phenotypes, favoring extremes; (C), stabilizing selection, weeds out extreme phenotypes allowing the majority of members within the population to converge on the intermediate physical feature (i.e., very light and very dark fur color may be less advantageous than a grayish, intermediate color).

17. D Just like the alternation of generations in plant life cycles may help to weed out unfit haploid genotypes, inbreeding depression (as it is called when the fitness of a population declines due to inbreeding) may wipe out the most unfit alleles, allowing the population to rebuild with a stronger overall gene pool. Inbreeding, however, is generally bad for a population because it greatly decreases the genetic variability of the population, making its members susceptible to even minor changes in the environment.

18. B Analogous traits are defined as characteristics that have a similar structure and function in two different organisms that are not evolutionarily related. Choice (A), homologous structures, refers to traits similar in structure because they derive from common ancestry (e.g., a chicken's wing and a human hand).

19. C The characteristics mentioned in the question stem describe a temperate deciduous forest, such as that which exists in the northern United States. These biomes experience a spring and summer growing season, cold winters, and plenty of moisture. All other answer choices are incorrect as they include biomes with climates and seasonal variability (markedly different) from the biome described in the question.

20. D Although moist skin is necessary for diffusion of nutrients and gases, it was not a novel (new) trait for land animals. Land animals do struggle to maintain moist membranes in their lungs and on their body surface; yet the gills of fishes and mollusks (as well as the tissues of a hydra) must be kept

moist for effective diffusion. All other traits mentioned in the answer choices describe novel adaptations to life on land that had not previously been present in aquatic life.

21. C This is a fairly difficult question with some trick answer choices. You might have been tempted to choose (D) if you simply added the frequency of individuals expressing blood type r (0.24 + 0.01) and did the same for blood type F (0.24 + 0.60). This type of addition works for the recessive r allele, but it does *not* work for the F allele. The frequency of the r trait, q^2, is ($rF + rf$), or 0.24 + 0.01 = 0.25; thus, the frequency of the r allele, q, is 0.5. However, individuals with blood type F can be either homozygous or heterozygous; therefore, the frequency of RF individuals in the question includes both FF individuals as well as Ff ones. So you cannot simply add 0.60, the RF frequency, to the rF frequency to calculate the frequency of the F allele. You must first find the frequency of the recessive f allele and then subtract that from 1.00 in order to find the frequency of the F allele. The frequency of the f trait is 0.15 + 0.01 = 0.16 so the frequency of the f allele is 0.4. The allele frequency of F is 1.00 − freq(f) = 0.60. So choice (C) is correct.

22. B Parthenogenesis is the production of offspring from unfertilized eggs. Offspring produced in this manner are usually haploid and form their own eggs by mitosis, not by meiosis. While parthenogenesis is an asexual form of reproduction, hermaphroditism is a sexual form of reproduction. Even though both gametes can, though rarely do, come from the same parent, the sperm and egg cells have each been produced in separate meiotic events. Therefore, hermaphroditism involves the genetic contributions of two parents, even when a hermaphrodite mates with itself. All other answer choices are incorrect regarding differences between these modes of reproduction.

GRE Biology Practice Test

1. Which of the following INCORRECTLY pairs a metabolic process with its site of occurrence?

   A. Glycolysis—cytosol
   B. Citric acid cycle—mitochondrial membrane
   C. Electron transport chain—mitochondrial membrane
   D. ATP phosphorylation—mitochondria
   E. Oxidative decarboxylation of pyruvate—mitochondria

2. Consider a biochemical reaction A → B, which is catalyzed by the human enzyme AB dehydrogenase. Which of the following statements is true of this reaction?

   A. The reaction will proceed until the enzyme concentration decreases.
   B. The reaction will be more favorable at 0°C.
   C. A component of the enzyme is transferred from A to B.
   D. The free energy change (DG) of the catalyzed reaction is the same as the free energy change for the uncatalyzed reaction.
   E. AB dehydrogenase will change the equilibrium but not the rate of the reaction to form A from B.

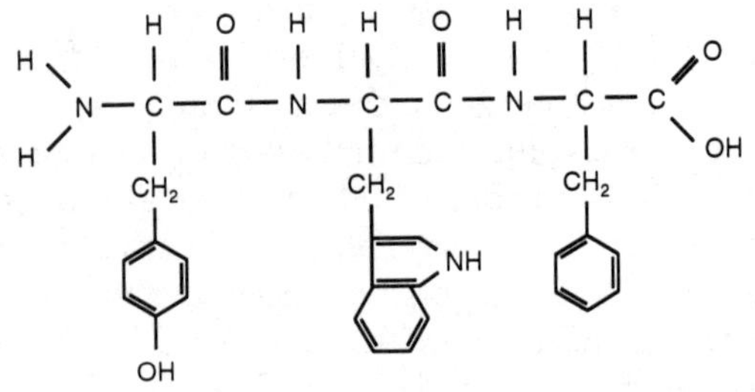

Figure 1

3. The molecule pictured above can be considered a(n):

   A. oligonucleotide.
   B. triglyceride.
   C. amino acid.
   D. trisaccharide.
   E. polypeptide.

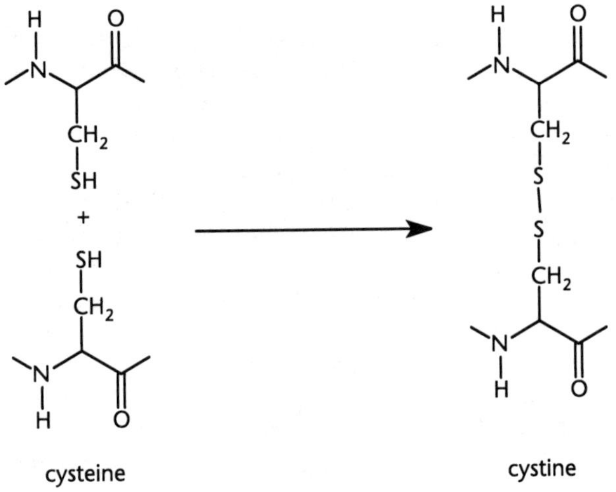

cysteine   cystine

Figure 2

4. The diagram above depicts a common bond formed between two nearby cysteine amino acids within a protein. The two cysteine molecules become oxidized to form cystine. The bond formed between them, consisting of two covalently bonded sulfur atoms, would most likely play a role in a protein's:

   A. primary structure.
   B. secondary structure.
   C. tertiary structure.
   D. quaternary structure.
   E. enzymatic capabilities.

5. The cancer drug Taxol interferes with the separation of chromosomes during anaphase of mitosis. The likely mechanism of Taxol's action is:

   A. interference with the synthesis of cyclin-Cdk complexes.
   B. prevention of kinetochore microtubule breakdown.
   C. excitation of inhibitory transcription factors.
   D. methylation of select areas of DNA on metaphase chromosomes.
   E. blockage of spindle formation from microtubule organizing centers.

6. In tabby cats, black coloration is caused by the presence of a particular allele on the X chromosome. A different allele at the same locus can result in orange color. The heterozygote calico cat, however, with splotches of both black fur and orange fur, is due to:

    A. polar body formation and inactivation.
    B. crossing over during gamete formation.
    C. the formation of thymine dimers.
    D. segregation of alleles in different cell lines.
    E. Barr body formation.

7. The DNA of a mouse is analyzed to reveal the presence of alleles resulting in brown fur color. The mouse is heterozygous, Bb, yet is completely colorless in phenotype. This can be explained as an occurrence of:

    A. epigenesis.
    B. epistasis.
    C. incomplete dominance.
    D. pleiotropy.
    E. codominance.

8. On a chromosome, genes A and W are five map units apart, W and G are 15 units apart, E and G are 12 units apart, and W and E are three units apart. Between which two genes would you expect the highest recombination frequency?

    A. A and W
    B. W and G
    C. E and G
    D. W and E
    E. A and G

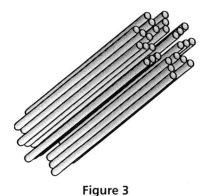

Figure 3

9. The preceding picture shows the arrangement of microtubules that can be found within:

    A. cilia.
    B. flagella.
    C. basal bodies.
    D. sarcomeres.
    E. motor kinesins.

10. Calmodulin is an intracellular signaling molecule that complexes with calcium ions in order to activate certain protein kinases. These protein kinases can be activated by the direct binding of Ca^{2+}/calmodulin complexes or by autophosphorylation of the kinase subunit. In the unphosphorylated state, one would expect the kinase subunit to:

    A. be constitutively active.
    B. phosphorylate target proteins that are nearby.
    C. be inactive.
    D. be unable to bind calmodulin.
    E. act as a transducer of calcium signals.

11. The sex-determination pathway in *Drosophila* results in the formation of either a male or female because of the presence or absence of key proteins coded for by a single gene known as *double-sex*. The *double-sex* gene codes for the protein dsx^f in order to build a female fruit fly, but it codes for the related protein dsx^m in order to build a male fruit fly. It is therefore appropriate to conclude that:

    A. the two protein products of the *double-sex* gene are created by alternative RNA splicing in male and female fruit flies.
    B. the protein products differ in males and females because females have two copies of the *double-sex* gene and males only have one.
    C. RNA polymerase in females reads the *double-sex* gene differently than it does in males, altering the transcription of the sex-determining gene.
    D. the dsx^f and dsx^m proteins must have different signal sequences, allowing them to go through alternative post-translational modifications.
    E. the proteins coded for by the *double-sex* gene have little to do with the presence of X or Y chromosomes in the cells of *Drosophila*.

12. Which of the following compounds are both products of the light reaction in photosynthesis and reactants in the Calvin cycle?

    A. Pyruvate and acetyl-CoA
    B. NADPH and O_2
    C. $NADP^+$ and CO_2
    D. $NADP^+$, ATP, and CO_2
    E. NADPH and ATP

13. The primary function of fermentation is to:

    A. generate ATP for the cell.
    B. synthesize glucose.
    C. regenerate NAD^+.
    D. synthesize ethanol or lactic acid.
    E. add to the total amount of ATP produced by cellular respiration.

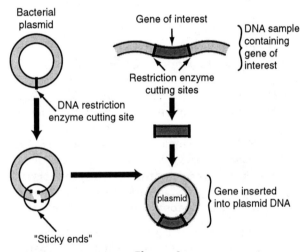

Figure 4

14. The preceding diagram shows the process used to insert a foreign gene into a bacterial plasmid, thereby transforming the bacteria into an organism that produces proteins typically produced in other organisms. This technique may fail unless the scientist:

    A. uses the entire mammalian stretch of DNA that contains the gene of interest, including the introns and exons that are part of that DNA.
    B. uses cDNA made from an RNA template of the gene of interest.
    C. inserts a reporter gene along with the gene of interest in order to measure the success of transformation.
    D. selects for antibiotic resistance before mixing the transformed plasmid with the bacterial cells.
    E. uses RNA nucleotides that can be directly converted into protein by the bacteria without first being transcribed.

15. All of the following are considered post-transcriptional modifications that occur in the nucleus EXCEPT:

    A. 5′ capping with methylated guanines.
    B. the addition of a 3′ poly-adenine tail.
    C. the excision of introns from mRNA via spliceosome formation.
    D. mRNA attachment to polyribosomes.
    E. stabilization of mRNA by snRNPs.

16. "DNA fingerprinting" uses RFLPs and gel electrophoresis in order to map the genetic differences between two individuals based upon:

    A. residue left from their fingerprints at a crime scene.
    B. slight differences in the lengths of pieces of their DNA after restriction enzyme digestion.
    C. single-nucleotide differences within genes for similar traits.
    D. differential gene expression within the cells of the two individuals.
    E. the existence of tissue-specific promoter sequences within short segments of their DNA.

17. In the cross between two individuals heterozygous for two different traits, it is expected that a 9:3:3:1 ratio of phenotypes will occur in the offspring. Yet, a ratio of 6:6:2:2 might indicate:

    A. pleiotropy.
    B. genomic imprinting.
    C. incomplete dominance.
    D. X-linkage.
    E. gene linkage.

Figure 5

18. In the diagram pictured above, the letter X represents:

    A. glucose.
    B. $NADP^+$.
    C. ATP.
    D. ADP.
    E. NADH.

19. Rotifers are tiny, pseudocoelomate animals that have complete digestive systems and other specialized organ systems. Which of the following statements are also true of rotifers?

    I. They possess both a mouth and an anus.
    II. Their internal body cavity is entirely enclosed by mesoderm.
    III. Their dorsal nerve cord runs superior to their notochord.

    A. I only
    B. II only
    C. II and III only
    D. I and III only
    E. I, II, and III

20. Which of the following would be LEAST able to make bacteria more virulent and infectious?

    A. The ability to evade nonspecific and specific body defenses
    B. The secretion of lipoteichoic acid, which contributes to septic shock
    C. Increasing the rate of bacterial replication
    D. Increasing the production of proteins allowing bacterial pili to extend and contact other pili
    E. The sudden ability of the bacteria to infect several different kinds of species rather than only one

21. Nitrogen-fixation can be carried out by free-living bacteria or by bacterial symbionts living in the roots of plants. The bacteria *E. coli* can use nitrate as an electron acceptor in the electron transport chain that is part of its cell membrane. The nitrate, therefore, serves the same purpose as which molecule in eukaryotes?

    A. Oxygen
    B. NADH
    C. Carbon dioxide
    D. Hydrogen
    E. NAD^+

22. The average distance between genes in *E. coli* is only 120 base pairs. Scientists have assigned functions to approximately 3,500 of the 4,000 genes that exist in this organism. Which of the following is also true of the *E. coli* genome?

    A. The coding sequences are continuous and lack noncoding introns.
    B. The genome consists of a series of linear chromosomes.
    C. RNA genes would be nonexistent as RNA is built solely at the ribosomes.
    D. There are multiple origins of DNA replication built into the genome.
    E. The genome contains a centromeric region necessary for chromosome segregation.

23. All of the following statements about transmission along neurons are correct EXCEPT:

    A. the rate of transmission of a nerve impulse is directly related to the diameter of the axon.
    B. the intensity of a nerve impulse is directly related to the size of the voltage change.
    C. a stimulus that affects the nerve cell membrane's permeability to ions can either depolarize or hyperpolarize the membrane.
    D. once initiated, local threshold depolarization stimulates the propagation of an action potential down the axon.
    E. the resting potential of a neuron is maintained by differential ion permeabilities and by the Na-K ATPase pumps.

24. Neurotransmitters characterized as inhibitory would not be expected to:

    A. open K^+ channels.
    B. open Na^+ channels.
    C. bind to receptor sites on the postsynaptic membrane.
    D. open Cl^- channels.
    E. hyperpolarize the neuron membrane.

25. The molecule pictured below could be best described as:

Figure 6

    A. a building block of a protein.
    B. an amino acid.
    C. a fatty acid.
    D. a waste product of amino acid metabolism.
    E. the amino portion of an amino acid.

26. The production of nitric oxide (NO) in the body is regulated by a complex of genes that includes the bNOS isoform. A knock-out mouse with a mutant bNOS gene was generated by recombinant molecular techniques. The mutant mouse produced normal bNOS protein except for the identity of amino acid 675 in the protein. Here, the amino acid cysteine was replaced with the amino acid tryptophan, rendering the bNOS protein nonfunctional. Which of the following mutations was responsible for this change?

    A. Frameshift
    B. Single base-pair deletion
    C. Point
    D. Nonsense
    E. Antisense

27. Sarcomere shortening in muscle fibrils requires:

    A. the rapid influx of sodium ions into the cytoplasm after release from the sarcoplasmic reticulum.
    B. T-tubule shortening after calcium ion release by the sarcoplasmic reticulum.
    C. ATP release from the sarcoplasmic reticulum and subsequent myosin attachment to actin filaments.
    D. conformational modifications of the tropomyosin-troponin complex within muscle fibers.
    E. an increase in the transcription of muscle regulatory proteins such as troponin and actin.

28. Which of the following is an example of passive immunity?

    A. A nurse gets stuck with a needle containing blood from a patient infected with tuberculosis (TB), gets a brief flu-like illness, and a few years later tests positive for anti-TB antibodies.
    B. A child receives a vaccination for polio consisting of inactivated polio virus.
    C. An adult exposed to a certain influenza strain will not become sick again because he was exposed to that strain as a child.
    D. A baby born to a woman who has antibodies to hepatitis may be temporarily resistant to the disease.
    E. A farmer exposed to a virus infecting some of his farm animals does not get ill when exposed to a related virus infecting his family.

29. Which of the following contribute(s) to gas exchange in the alveoli?

   I. Low partial pressure of O_2 in the pulmonary capillaries when compared to the inhaled air
   II. Low partial pressure of CO_2 in the pulmonary capillaries when compared to inhaled air
   III. Presence of surfactant

   A. I only
   B. II only
   C. II and III only
   D. I and III only
   E. I, II, and III

30. All of the following can explain why the removal or rearrangement of groups of cells at pregastrula stages in vertebrate embryos can be tolerated by the embryos, and allow the embryos to develop normally, EXCEPT:

   A. destruction of small numbers of cells is not harmful because other cells nearby can replace them.
   B. rearrangement of cells within a group has no effect because all cells in the pre-gastrula stage are equivalent in competence.
   C. moving cells to other areas of the embryo is tolerated because cells can adapt to their surroundings without being in the presence of an inducer.
   D. moving cells that will become organizing inducers to other regions will simply allow cells in that new location to respond according to their distance from the inducer.
   E. the selective gene loss occurring within the pregastrula cells allows certain groups of cells to be removed without detrimental effect.

31. Lichens are a mutualistic combination of which of the following organisms?

   A. Fungus and bacteria
   B. Bacteria and algae
   C. Green plant and fungus
   D. Bacteria and green plant
   E. Fungus and algae

32. The haploid structures of fungi through which sexual reproduction occurs are known as:

   A. gametangia.
   B. mycelia.
   C. haustoria.
   D. mycorrhizae.
   E. dikaryotic hyphae.

33. Which of the following characteristics could most readily distinguish a marsupial from a placental mammal?

   A. Egg-laying ability
   B. Fusion of the lower jaw bones in the marsupial
   C. The presence of a layer of insulating hair in the placental
   D. Whether or not the young are nourished by their mother's milk
   E. The degree of fetal development at the time of birth

34. Members of the phylum Bacillariophyta are yellow or brown eukaryotic algae. They have unique, glasslike walls made of silica and the walls are built in two sections that fit together like a box lid and a box. They are also known as:

   A. dinoflagellates.
   B. zygomycetes.
   C. diatoms.
   D. green algae.
   E. volvox.

35. A bilaterally symmetric deuterostome might be classified in the phylum:

   A. Annelida.
   B. Chordata.
   C. Platyhelminthes.
   D. Porifera.
   E. Arthropoda.

36. Which of the following statements is not consistent with Darwin's theory of natural selection?

    A. Individuals in a population exhibit variations, some of which can be passed along to offspring.
    B. Organisms change during their life-spans to better fit their environment, and these changes can be passed along to offspring.
    C. Natural selection can lead to speciation.
    D. Individuals that reproduce most successfully are more likely to have offspring that also reproduce successfully if the environment remains stable.
    E. Certain organisms have a higher rate of reproductive success than other organisms due to a variety of environmental factors.

37. A community is made up of:

    A. the factors that create an organism's niche.
    B. the interaction of several ecosystems.
    C. populations of organisms and their abiotic environment.
    D. several different populations of organisms living together in the same habitat.
    E. a group of organisms that interact in a mutualistic fashion.

38. Two plants growing together in the same pot are separated and planted in different pots. One plant dies while the other grows much higher. The plants most likely had what kind of relationship?

    A. Parasitic
    B. Commensalism
    C. Mutualistic
    D. Mycorrhizal
    E. Competitive

39. A hormone able to exert negative feedback on both the hypothalamus and pituitary glands during the menstrual cycle is:

    A. FSH.
    B. LH.
    C. hCG.
    D. GnRH.
    E. estrogen.

40. A major difference between ectotherms and endotherms is that:

    A. as ambient temperature rises, ectotherms maintain nearly constant body temperature.
    B. endotherms receive most of their body heat from their surroundings.
    C. endotherms derive body heat from metabolic reactions and use energy derived from metabolic reactions to cool their bodies.
    D. ectotherms maintain their body at lower temperatures than do endotherms, therefore leading to the term "cold-blooded."
    E. ectotherms cannot live on land because of temperature fluctuations that can damage their organ systems.

41. Benefits of asexual reproduction include all of the following EXCEPT:

    A. it often allows for the production of many more offspring at the same time.
    B. it is advantageous in changing environments in which population variety is the key to successful propagation of a species.
    C. it is easier in certain environments to have offspring without searching for a mate.
    D. structuring the social organization of certain species of social insects, in which certain members of the species are produced asexually through parthenogenesis.
    E. allowing the conservation of resources otherwise allocated to finding mates and performing ritualized courtship.

42. The bark of a tree is made up mainly of:

    A. xylem.
    B. pulp.
    C. vascular cambium.
    D. tracheid cells.
    E. phloem.

43. Which of the following is true regarding phage lambda, a virus that infects bacteria?

    A. In the lytic cycle, the bacterial host replicates viral DNA, passing it on to daughter cells during binary fission.
    B. In the lysogenic cycle, the bacterial host replicates viral DNA, passing it on to daughter cells during binary fission.
    C. In the lytic cycle, viral DNA is integrated into the host genome.
    D. In the lysogenic cycle, the host bacterial cell bursts, releasing phages.
    E. Phage lambda is able to replicate without entering a bacterial host.

44. The covalent binding of a molecule other than the substrate to the active site of an enzyme often results in:

    A. noncompetitive inhibition.
    B. feedback inhibition.
    C. irreversible inhibition.
    D. increasing the energy of activation.
    E. modifying the free energy change of the reaction.

45. In a food chain that consists of grass → grasshoppers → spiders → mice → snakes → hawks, the organism(s) that possess the most biomass within the community is (are) the:

    A. grass.
    B. grasshoppers.
    C. mice.
    D. snakes.
    E. hawks.

46. Avascular plants that alternate between haploid and diploid forms and have flagellated sperm are members of the division:

    A. Rhizopoda.
    B. Bryophyta.
    C. Pterophyta.
    D. Coniferophyta.
    E. Anthophyta (the angiosperms).

47. Which of the following is not a characteristic of monocots?

    A. One cotyledon in each seed
    B. Parallel leaf veins
    C. Fibrous root system
    D. Vascular bundles of xylem and phloem complexly arranged
    E. Petals in multiples of four or five

48. Some members of this phylum of invertebrates undergo torsion during early embryonic development, resulting in a mantle cavity and anus that are above the head in adults:

    A. Annelida
    B. Rotifera
    C. Chordata
    D. Mollusca
    E. Cephalochordata

49. The punctuated equilibrium hypothesis claims that:

    A. cataclysmic events (e.g., asteroid strikes) have shaped the history of life on earth.
    B. most speciation occurs sympatrically.
    C. species go through long periods of time during which they do not change markedly in genotype or phenotype.
    D. new species arise through mutations that have large effects on phenotype.
    E. the boundaries between fossil species are largely arbitrary.

50. Use the following data regarding crossover frequencies to map the relative location of four genes linked on the same chromosome: P, Q, R, and T.

Table 1

Genes	Frequency of Crossover
P and Q	35%
R and Q	20%
R and P	15%
T and Q	60%
P and T	25%

Which of the following represents a possible arrangement of genes P, Q, R, and T on the chromosome where they are linked?

A. P-Q-R-T
B. P-R-Q-T
C. T-P-R-Q
D. T-P-Q-R
E. Q-P-T-R

51. In some organisms, features that have no function become vestigial and are ultimately lost. In many cave-dwelling animals, organs such as the eyes have been lost while other sense organs have increased in size. Which of the following hypotheses to explain the loss of nonfunctioning organs would not be considered correct?

A. Mutations causing the reduction in size of nonfunctional traits become fixed by genetic drift.
B. Natural selection against organs that are not used exists because the organs interfere with other, more important, functions.
C. The development of the organ requires energy expenditures that are better spent on building other tissues or maintaining other traits.
D. All organs are maintained or eliminated as a result of how much they are used.
E. The organs that disappear have a negative genetic correlation with other traits.

52. An iteroparous life history, rather than a semelparous one, may evolve in a certain species if:

A. juvenile individuals devote a great deal of energy toward reproductive efforts.
B. environmental conditions necessitate having a large batch of offspring early on in life.
C. young individuals devote much energy early on to self-maintenance, while saving reproductive efforts for later on in life.
D. delayed maturation increases the chances of dying prior to reproducing.
E. there is a high adult mortality rate in its particular habitat.

53. Male birds that leave their mates immediately after mating and go off in search of other females often benefit by fathering so many offspring. This is known as:

A. polygamy.
B. monogamy.
C. polygyny.
D. polyandry.
E. altruism.

54. Heterochrony is defined as an evolutionary change in the rate of development of a particular feature. In some cases, a feature present in adult members of a species resembles the same feature that was present in juveniles of the ancestors of that species. These underdeveloped adult characteristics are often called:

A. ontogenetic.
B. morphogenetic.
C. paedomorphic.
D. heterotropic.
E. homeotic.

55. Aneuploidy is the most common type of chromosomal disorder. The condition may arise from either a trisomy or a monosomy. The only monosomy that is generally nonlethal is known as:

A. Turner syndrome.
B. Klinefelter syndrome.
C. XYY syndrome.
D. Down syndrome.
E. Cri-du-chat syndrome.

56. All of the following statements are true of the hormone estrogen EXCEPT:

    A. estrogen is a steroid hormone.
    B. spikes in estrogen concentration in the blood bring about surges in luteinizing hormone, which causes ovulation.
    C. estrogen is related to testosterone in chemical structure.
    D. estrogen is released by the cells in the ovaries and acts only locally within the developing follicles.
    E. estrogen and progesterone are secreted by the corpus luteum in order to maintain the thickness of the endometrium.

57. When a muscle fiber is subjected to very frequent stimuli:

    A. an oxygen debt is incurred.
    B. a muscle tonus is generated.
    C. the threshold value is reached.
    D. the contractions combine in a process known as summation.
    E. a simple twitch is repeatedly generated.

58. All of the following are examples of imperfections in the fossil record that make it difficult to distinguish patterns such as phyletic gradualism and punctuated equilibrium EXCEPT:

    A. ages of most fossils can be estimated only in an imprecise manner, since most fossils themselves cannot be dated directly by radiometric techniques.
    B. fossils deposited over a short time interval are usually mixed together with other fossils before the sediment solidifies, such that a sample of fossils is often a time-averaged sample.
    C. paleozoic insect fossils are usually flattened imprints so that many character traits cannot be studied.
    D. mesozoic mammals often leave only a jawbone or tooth behind as their only trace.
    E. fossils of early Foraminifera show that the species had evolved rapidly from one relatively stable phenotype to another at the Miocene/Pliocene boundary.

59. Growth rings of plants, studied by dendrochronologists, consist of which of the following tissue(s)?

    A. Leaf mesophyll
    B. Pericycle and stele
    C. Secondary xylem and phloem
    D. Protoderm and procambium
    E. Apical meristem

60. What function do lenticels, pneumatophores, and root hairs all have in common?

    A. They are used to absorb nutrients in the form of nitrogen- and phosphorus-bearing compounds.
    B. They are used to anchor plants into the ground.
    C. They are used for gas exchange, often in plants whose roots are underwater.
    D. They are involved in the maintenance of hydrostatic pressure at the roots for proper conduction of sugar within the phloem.
    E. They are responsible for forming mycorrhizal associations with bacteria and fungi.

61. Restoration of the resting state in muscles begins when neural stimulation stops and calcium ions are transported back into the:

    A. postsynaptic terminal of the nearby axon.
    B. sarcoplasmic reticulum.
    C. presynaptic axon terminal.
    D. neuromuscular junction.
    E. T-tubules.

62. Two plants, X and Y, are grown as potential food crops. Plant X is able to maintain a high rate of photosynthesis as oxygen level in the air around it increases from a low of 10% to a high of 50%, yet plant Y's rate of photosynthesis drops drastically under these circumstances. The best conclusion to draw from this data is that:

    A. plant X is a CAM plant.
    B. plant Y is performing only the Calvin cycle in higher oxygen partial pressures.
    C. plant X is a CAM plant and plant Y is a C_4 plant.
    D. plant Y is performing only the light reactions of photosynthesis.
    E. plant X is a C_4 plant, and plant Y is a C_3 plant.

63. Which of the following terms or phrases would not be associated directly with photosystem II in plants?

    A. Photophosphorylation
    B. The splitting of water
    C. Harvesting light energy by chlorophyll
    D. Oxygen released from water
    E. Chlorophyll a

64. All of the following are examples that are evidence of the endosymbiont theory EXCEPT:

    A. mitochondria and chloroplasts are both affected by drugs that halt protein synthesis in prokaryotes.
    B. mitochondrial inner membranes are similar in structure to bacterial cell membranes.
    C. some modern-day bacteria can live outside eukaryotic cells.
    D. fossils carbon-dated to over 500 mya show prokaryotes lived within eukaryotic cells.
    E. both mitochondria and chloroplasts possess their own DNA and can reproduce independently of the cell they are within.

65. The water-soluble vitamin that acts as a co-enzyme in nucleic acid metabolism and can be found in vegetables and grains is:

    A. vitamin B_6.
    B. niacin.
    C. folic acid.
    D. vitamin A.
    E. ascorbic acid.

Questions 66–68

    A. Endosperm
    B. Gametophyte
    C. Protoderm
    D. Pericarp
    E. Cotyledon

66. Embryonic seed leaf that has a large surface area to absorb nutrients during seed germination

67. The thickened wall of the fruit, derived from the ovary

68. Multicellular haploid form of a plant that undergoes mitosis to produce haploid gametes

KAPLAN
Test Prep and Admissions

GRE Biology Practice Test: Answers and Explanations

ANSWER KEY

1. B (See Concepts 9.2, 9.3, and 9.4)
2. D (See Concept 8.4)
3. E (See Concept 5.4)
4. C (See Concept 5.4)
5. B (See Concept 12.3, pp. 232–233)
6. E (See Concept 15.3, p. 284)
7. B (See Concept 14.3)
8. E (See Concept 15.2, pp. 279–281)
9. C (See Concept 6.6)
10. C (See Concept 11.3, p. 1072)
11. A (See Concept 17.3)
12. E (See Concepts 10.2 and 10.3)
13. C (See Concept 9.5)
14. B (See Concept 20.1, pp. 387–388)
15. D (See Concept 17.3)
16. B (See Concept 20.5, pp. 404–405)
17. E (See Concept 15.2)
18. C (See Concepts 9.3 and 9.4)
19. A (See Concept 33.3, pp. 648–649)
20. D (See Concept 18.3)
21. A (See Concept 9.4)
22. A (See Concept 18.3)
23. B (See Concept 48.3)
24. B (See Concept 48.4, pp. 1024–1025)
25. D (See Concept 44.2)
26. C (See Concept 17.7)
27. D (See Concept 49.6)
28. D (See Concept 43.3, p. 914)
29. D (See Concept 42.7)
30. E (See Concept 47.1, pp. 992–997)
31. E (See Concept 31.5, pp. 621–622)
32. A (See Concept 31.2)
33. E (See Concept 34.7)
34. C (See Concept 28.5, pp. 559–560)
35. B (See Concept 32.3)
36. B (See Concept 22.2)
37. D (See Chapter 53, p. 1159)
38. A (See Concept 53.1)
39. E (See Concept 46.4)
40. C (See Concept 40.5, pp. 833–834)
41. B (See Concept 13.1, p. 239)
42. E (See Concept 35.4)
43. B (See Concept 18.1, pp. 337–339)
44. C (See Concept 8.5)
45. A (See Concept 54.3, pp. 1192–1193)
46. B (See Concept 29.3)
47. E (See Concept 30.3, pp. 602–603)
48. D (See Concept 33.4)
49. C (See Concept 24.2, pp. 481–482)
50. C (See Concept 15.2, pp. 279–281)
51. D (See Concept 22.3, pp. 448–449)
52. C (See Concept 52.2)
53. A (See Concept 51.5, pp. 1123–1127)
54. C (See Concept 24.3, pp. 484–485)
55. A (See Concept 15.4)
56. E (See Concept 46.4)
57. D (See Concept 49.6)
58. E (See Concept 25.1, p. 492)
59. C (See Concept 35.4)
60. C (See Concept 35.1, pp. 713–714)
61. B (See Concept 49.6)
62. E (See Concept 10.4)
63. A (See Concept 10.2, pp. 189–193)
64. C (See Concept 26.4)
65. C (See Concept 41.2, p. 851)
66. E (See Concept 30.3, pp. 598–600)
67. D (See Concept 30.3, pp. 598–600)
68. B (See Concept 30.3, pp. 598–600)

EXPLANATIONS

1. B The citric acid cycle, otherwise known as the Krebs cycle, takes place within the matrix (or inner fluid) of the mitochondria. All other answer choices are correctly placed within their proper areas of occurrence.

2. D Although enzymes may decrease the activation energy needed to start a reaction, the overall free energy change between products and reactants does not change. In other words, products and reactants maintain the same potential energy difference in uncatalyzed reactions as they do in catalyzed ones. Be careful of (E) since this statement is the opposite of what enzymes do.

3. E The molecule pictured here is a sequence of three amino acids linked together by peptide bonds. When trying to figure out what molecule this might be, the giveaway is the series of N-C-C bonds that occur. Amino acid chains all have this sequence with R groups hanging off the middle C of each amino acid. Saccharides are sugars, while glycerides are lipids, so those answer choices should be eliminated.

4. C The bond shown here is a disulfide bridge, formed between cysteines that are across from each other in a protein's tertiary structure. The bonds help anchor and hold the 3-D tertiary structure together. Keep in mind that the primary structure of a protein is simply the amino acid sequence, while the secondary structure is made up of β-pleated sheets and α helices caused by hydrogen-bonding between nearby amino acid side chains. Covalent bonds, such as disulfide bridges, make up tertiary structure only.

5. B Chromosomes separate to opposite poles of the cell during anaphase of mitosis due to the rapid shortening of kinetochore microtubules, those parts of the spindle that are attached to the centromere regions of the chromosomes. Taxol interferes with the breakdown of these microtubules, thereby halting the separation of the chromosomes and stopping mitosis in its tracks, perfect for halting the spread of a tumor. No new transcription is required at this point (C and D are wrong) and spindle formation occurs earlier on in mitosis.

6. E Barr bodies are inactivated X chromosomes commonly found in cells of female mammals. One of the two X chromosomes in each cell within the organism condenses to the side of the nucleus and cannot be used for transcription. The X chromosome of each pair that inactivates is apparently random, and results in some cells expressing traits coded for by alleles found on one X chromosome, while other cells express traits coded for by alleles on the other X. This leads certain patches of calico tabby fur to express orange coloration while other patches remain black.

7. B Epistasis occurs when one gene alters the expression of another gene that is independently inherited. In this instance, the brown fur color allele is most likely affected by the presence of recessive alleles on a different chromosome that code for the expression of "no color." Don't confuse epistasis with epigenesis, the phenomenon where genes are expressed differently depending upon which parent they are inherited from.

8. E If you were to draw out the chromosome and map the genes to this piece of DNA, you would see that they are in the following order (shown with map units between them): A—5—W—3—E—12—G. Thus, genes A and G are furthest from each other and most likely to split apart during a crossing over

event in meiosis. A and G have the highest frequency of recombination. Mapping genes to a chromosome is like a puzzle—you have to try various positions of the genes until they fit together linearly in such a way that their distances from each other all work out according to the information given.

9. C The structure pictured, showing nine microtubule triplets, can be found in basal bodies, which form the base of eukaryotic cilia and flagella. Within the cilia and flagella, however, are nine microtubule doublets surrounding a central pair of microtubules. Centrioles have the same structure as basal bodies, yet are often found as pairs oriented at 90-degree angles to one another within the cytoplasm.

10. C Proteins can be activated in cells via phosphorylation. Kinases are enzymes that phosphorylate other enzymes or proteins. Yet, being proteins themselves, the kinases must generally be phosphorylated in order to work. So in an unphosphorylated state, the kinase described in the question stem would be inactive. Calmodulin is the transducer (messenger) of intracellular calcium signals here.

11. A The key here is to recognize that the question describes two protein products that are both coded for by the same gene. In general, these differences come from the fact that various exons within a gene can be spliced together in several ways to allow for the production of many different, but related, proteins from a single stretch of DNA. In this case, the proteins coded for by *double-sex* are different because of alternative RNA splicing after transcription.

12. E The light reactions of photosynthesis produce electrons, which get picked up by $NADP^+$, as well as ATP due to a hydrogen ion gradient that is formed within the thylakoid membrane. The $NADP^+$ gets oxidized to NADPH as it gains electrons, and it is used along with the ATP to power the Calvin cycle, where sugars are produced. Don't be tempted by choice (B), since oxygen is a waste product of the light reactions, and is not used in the Calvin cycle.

13. C The key function of fermentation is to regenerate NAD^+. The reason this is so essential is that as glucose gets broken down within the cytoplasm of a cell, electrons fly out of the broken bonds. These electrons are normally caught by NAD^+ and ferried into the mitochondria and ETC. Yet, fermentation occurs in the absence of the mitochondria and ETC. Thus, in order to keep a ready supply of NAD⁺ around in order to catch electrons coming off the broken glucose molecules, the NADH must be re-oxidized into NAD⁺ by the process of fermentation. This allows the breakdown of glucose to continue and the subsequent production of small amounts of ATP via glycolysis.

14. B   The main problem with trying to insert a eukaryotic gene into a bacterium is that bacteria have no introns, and thus have no machinery to remove them. Choice (A) is incorrect here for that reason. Because of this, the most successful transformations can be achieved using cDNA, or complementary DNA, which is made of an mRNA template of the gene of interest. The mRNA has already been spliced to remove introns, so using that mRNA to create a complementary DNA molecule lets you use the "pure" gene without exons in your experiment. Although the use of reporter genes and antibiotic resistance may be important after the transformation occurs (in order to select transformed colonies), reporter genes are not needed for the success of the experiment (choice C) nor can one select for antibiotic resistance *before*hand.

15. D   All of the choices listed describe posttranscriptional modifications, those that take place *in the nucleus* after transcription, except choice (D), the attachment of mRNA to polyribosomes. Polyribosomes are groups of multiple ribosomes that can attach to mRNA to read the codons and produce multiple polypeptide chains at the same time. This takes place outside the nucleus during translation.

16. B   RFLPs are restriction fragment length polymorphisms. If two individuals' DNA is cut up using the same restriction enzyme, there will be differences in the lengths of the fragments created because the individuals have slight differences in nucleotide sequences within their DNA. These restriction fragments can be compared using a gel to separate the fragments by size. The banding patterns on the gels for each individual can be analyzed to determine where genetic differences may lie. RFLPs may be caused by single nucleotide differences, as in choice (C); however, DNA fingerprinting is not based on the analysis of single nucleotide differences between two individuals, but rather much larger sections of DNA.

17. E   Gene linkage leads to phenotypes in the offspring that are similar to the parental phenotypes. In fact, the more closely linked the genes are on the same chromosome, the less likely it is that recombinant (nonparental) phenotypes will occur due to crossing-over. Do not confuse gene linkage with X-linkage, which is simply when genes are located on the X-chromosome.

18. C   This is a diagram of a mitochondrion. It is here that the reactions of cellular respiration take place. Pyruvate, produced from the breakdown of glucose in glycolysis, enters the mitochondrial matrix along with oxygen and NADH, the Krebs cycle takes place (as well as electron transport and oxidative phosphorylation), and ATP is produced. In fact, the letter X represents ATP, whose production is the most essential aspect of the entire cellular respiration process.

19. A   Being invertebrates and not chordates, rotifers do not have a notochord or dorsal nerve cord. In addition, being pseudocoelomate means that their internal body cavity is only partially covered with mesoderm. The question stem states that rotifers have a complete digestive system, so that is equivalent to their having a mouth and an anus. Make sure to know the basic vocabulary of animal phyla for the test!

20. D   Of all the answer choices, the only one that does not deal with bacterial infection of other organisms' cells is choice (D). While the ability of bacterial cells to exchange genetic information through pili help antibiotic-resistance spread, all the other choices are more direct causes of increased virulence within a given organism. Look out for answer choices such as (D) that do not fit in with the pattern given by all the other choices.

21. A   Neither NADH nor NAD⁺ serve as electron acceptors within the electron transport chain of eukaryotes. However, oxygen does. Recall that oxygen is the terminal electron acceptor in the ETC of most organisms, allowing water to be formed from hydrogen and electrons coming off the ETC. Without the presence of oxygen, ATP production would rapidly stop.

22. A   *E. coli* is a bacterial species, and bacteria have no introns within their genes. This is in contrast to eukaryotes, which not only have introns but also possess multiple origins of replication (rather than *E. coli*'s single origin), and linear chromosomes (rather than the single, circular one of bacteria). Choice (C) must also be eliminated since RNA genes must be found on DNA if they are to be transcribed at all.

23. B Neurons send their signals in an all-or-none fashion, meaning that once they initiate an action potential, the internal voltage of the neuron climbs rapidly to a certain level no matter the scale of the initial stimulus. In other words, most neurons reach an internal voltage of +50 mV for each action potential no matter how intense the stimulus is. The reason that you can feel one stimulus as being more intense than another has to do with the number of different neurons involved or with the frequency of firing, not the voltage change of any individual neuron. All other statements regarding nerve cell potentials are correct.

24. B Neurotransmitters that are inhibitory would not be expected to open sodium channels, because the opening of sodium channels rapidly increases the flow of positive ions into the axon, resulting in depolarization. Again, if this depolarization is great enough, the axon's membrane will pass a threshold point and an action potential will be initiated. Inhibitory neurotransmitters will, however, open chloride ion or potassium ion channels, both of which will hyperpolarize the inside of the axon relative to the outside. Chloride channels will cause the inflow of negative chlorine ions, making the inside of the axon more negative and, thus, harder to depolarize; and the opening of potassium channels will cause the outflow of positive potassium ions, again causing the inside of the axon to become more negative and harder to depolarize.

25. D The molecule pictured is urea, a nitrogenous waste product formed as amino acids and proteins are broken down by cells. Urea picks up free amino groups (—NH$_2$) and links them via a carbon atom so that they can be filtered out by the kidney and eliminated in the urine. Remember that amino acids all have a carboxyl group (—COOH) as well and urea does not.

26. C The only mutation listed that could change one amino acid into another amino acid without altering the rest of the protein's primary sequence is a point mutation, changing one base out of the three that make up the codon coding for cyteine. A nonsense mutation, choice (D), would result in a change

that would cause a premature termination of the protein rather than replace one amino acid with another. Frameshift mutations, including single base-pair deletions would shift the entire reading frame of the mRNA strand from the point of mutation onward, thus changing many amino acids in the primary sequence.

27. D Recall that a sarcomere is a unit of layered actin and myosin filaments, and within a given muscle fiber there are repeating sarcomere units that traverse the length of the muscle cells. The shortening of the sarcomeres is regulated by a nerve impulse outside the fiber, resulting in calcium ion release from the sarcoplasmic reticulum, a specialized ER found in muscle cells. Calcium allows a conformational change to take place within the sarcomere such that tropomyosin is pulled off of the binding site (troponin) for the myosin heads on the actin filaments. With the myosin binding sites uncovered, myosin can attach to actin and slide past it, shrinking the length of the sarcomere.

28. D Passive immunity results when antibodies are transferred from one individual to another. Often this occurs when a pregnant woman passes on her own antibodies to her fetus through the placenta, or to her baby via breast milk. Since antibodies are proteins, they are eventually degraded, and passive immunity may last only for a few days or weeks. All choices other than (D) are examples of active immunity, whereby one's immune system, after being exposed to a particular pathogen, mounts a response and creates memory B and T lymphocytes that can be activated very quickly in the event of reexposure.

29. D Compared to inhaled air, the pulmonary capillaries have a higher partial pressure of CO$_2$ and a lower partial pressure of O$_2$. Thus, in the alveoli, gas exchange occurs when CO$_2$ flows down its concentration gradient from the capillaries to the alveoli, and O$_2$ flows down its gradient from the alveoli into the capillaries. The presence of surfactant lowers the surface tension of the alveoli and keeps the membranes moist to facilitate gas exchange. Therefore, statements I and III contribute to gas exchange—choice (D).

30. E Keep in mind for this question that pregastrula cells have not yet differentiated. In other words, the blastula cells in chordates have the potential to turn on or off any of their genes as gastrulation occurs and endo-, meso-, and ectoderm begins to form. Removal or movement of nondifferentiated cells, even cells that have yet to become organizers, cannot cause any effect because these cells have not yet differentiated. Selective gene loss would not occur in pregastrula cells and, if it did, it would result in certain cells already taking steps to develop along certain lines. The removal or rearrangement of these cells, then, could result in drastic changes in the embryo. In addition, selective gene loss generally does not occur in chordates.

31. E Lichens are symbiotic unions of a fungus and an alga. The fungus supplies water and other absorbed nutrients, while the alga supplies various products of photosynthesis. They live in many unpolluted areas, on rocks, bark, and on the ground.

32. A Gametangia are the separate hyphae, one male and one female, that contain several haploid nuclei each and reach toward each other to fuse, a process called plasmogamy. The fusion of these hyphae gives rise to dikaryotic hyphae, or hyphae that have two nuclei, which only later fuse to create a fully diploid cell. Mycorrhizae are mutualistic associations between fungi or plants and symbiotic bacteria that help fix nitrogen and other substances for use in the fungus or plant. Haustoria are parasitic hyphae that contain nutrient-absorbing tips to penetrate the host's cells.

33. E The only major difference between marsupials and placentals is that marsupials' young are born (leave the uterus) at a much earlier stage than the young of placentals. Marsupials finish out their development within a pouch filled with mammary glands for nourishment. Placentals also have mammary glands (although not in a pouch). All mammals have a layer of insulating hair and only the monotremes (such as the duck-billed platypus) lay eggs.

34. C The diatoms are algae known for their boxlike silica shells. Accumulations of these shells on the seafloor can be collected and used for many commercial purposes. You should have eliminated choice (B), zygomycetes, because the suffix *-mycetes* should signal to you that that is a fungus or fungus-related.

35. B Although all of the phyla listed as answer choices are bilaterally symmetric (except for Porifera, the sponges), only one has organisms classified as deuterostomes—Chordata. You should remember for the GRE that deuterostome embryos

have indeterminate cleavage. Each cell produced in the early zygote by mitotic cleavage retains the ability to become a complete embryo on its own. There is no selective gene loss in chordates. In contrast, the other phyla mentioned in the answer choices (again, with the exception of Porifera) are either protostomes or are not coelomate organisms (and, thus, are not classified as proto- or deuterostomes).

36. B Organisms may adapt and respond to their environment during their lifetime, yet these changes are not heritable (able to be passed down to offspring). Choice (B) is much more Lamarckian than Darwinian in nature, and it should be picked as the correct answer here. All other statements are Darwinian in content.

37. D A community is defined as a group of different species living together in the same habitat. The word community, as differentiated from the word ecosystem, does not include the non-living, or abiotic, factors within the habitat of these populations.

38. A If one plant dies while the other is strengthened because of this separation, it is likely that one plant was preying on the other as a parasite. Only after removal does the host species grow taller, but the parasite will die off because of lack of nutrition. It is not clear that the separation would affect plants in a commensalistic relationship, and those in a competitive relationship would both be strengthened by the separation.

39. E Along with progesterone, estrogen is able to exert a negative feedback on the brain, blocking the secretion of both LH and FSH in the anterior pituitary and GrR4 in the hypothalamus. However, *high* concentrations of estrogen, released prior to ovulation, cause a sudden surge in LH secretion by the anterior pituitary that results in ovulation. In addition, the corpus luteum is maintained by high LH concentration and is responsible for its own destruction because it secretes estrogen and progesterone that feedback inhibit the pituitary. As soon as estrogen and progesterone levels rise, approximately 10–12 days after the corpus luteum forms, LH concentration drops and the corpus luteum breaks apart.

40. C The major difference between endo- and ectotherms comes not from the temperature at which either type of organism maintains its body, but rather from the source of body heat that each type of organism uses. Ectotherms derive most of their body heat from the surrounding environment—surrounding temperature is termed the "ambient" temperature. However, endotherms can generate body heat or cool themselves off using metabolic reactions to keep their body temperature fairly constant.

41. B All of these statements concerning asexual reproduction are correct, except that asexual reproduction is best in favorable, stable environments, ones that don't change rapidly. The reason for this is that asexual reproduction, in contrast to its sexual counterpart, results in the formation of identical offspring. Although asexual organisms can often produce many more offspring in a single reproductive event than sexual organisms, these asexually produced young do not usually have the genetic variation caused by meiosis and crossing-over to be able to survive a rapidly changing environment or times of environmental stress.

42. E Bark is made up of phloem and periderm. The periderm is a tissue built from a cylinder of meristem called cork cambium, which can be found in the outer parts of the stem. Xylem, made up of tracheids and vessel elements, is found internally. The reason that trees will die if you cut off their bark is that they lose the means by which to ferry sugars down from the leaves to the roots, since their phloem tissue has been destroyed.

43. B Phage lambda must enter a bacterial host to replicate. The phage has two methods of reproduction: the lytic cycle and the lysogenic cycle. In the lytic cycle, the phage attaches to a host bacterial cell and injects its DNA into the bacterium. The virus utilizes the nucleotides, enzymes, and ribosomes of the host bacterium to replicate, and it organizes replicated viral DNA as well as viral proteins to build new phages. The viral particles will kill the host bacterium as they burst out of its cell membrane. In a lysogenic cycle, the phage attaches to a host bacterial cell and injects its DNA into the cell, after which the viral DNA gets integrated into the bacterial genome, remaining there and copying itself over many generations as the bacterial cell replicates.

44. C In irreversible inhibition, a molecule other than the substrate covalently bonds to the active site of the enzyme, preventing substrate molecules from accessing the active site. Although competitive inhibition, where chemicals compete with substrate molecules for a spot within the active site, may slow down an enzymatic reaction, irreversible inhibition may stop a reaction altogether and permanently. Noncompetitive inhibition takes place when a chemical decreases the affinity of an enzyme for a substrate through binding to a location on the enzyme other than the active site, inducing a conformational change in the active site without ever entering that area.

45. A Organisms at the top of the food chain have the least biomass, while organisms at the bottom have the greatest biomass. In this food chain, grass is at the bottom and has the greatest biomass.

46. B Bryophytes (mosses) are avascular plants that alternate between haploid and diploid generations and have flagellated sperm. Make sure to know the basic differences between plant and animal phyla!

47. E Dicots, not monocots, have parts of their flowers (petals) in multiples of four or five; monocots, however, have petals in multiples of three. All other characteristics in the answer choices are traits of monocots.

48. D Members of the phyllum Mollusca (snails, for example) undergo torsion during development and possess a mantle and anus above their heads. You could also have taken a stab at the answer by recognizing that the word mantle should be associated with creatures such as molluscs.

49. C Punctuated equilibrium postulates that most evolutionary change happens quickly in small isolated populations that break off from larger groups, so this eliminates (B). Between bursts of change, phenotypes in a population hover around some mean value, not changing very much. Even though change is rapid in punctuated equilibrium, it relies on variation in the population, not on the sudden production of novel features. In the fossil record, two different phenotypes would mark one of these punctuation events and would, therefore, not be arbitrary.

50. C Two genes that are close together on a chromosome will have a low frequency of separating from one another during crossover events. The frequency that they do split up and end up in different gametes from each other is called the recombination

KAPLAN
Test Prep and Admissions

frequency. The closer two genes are, the lower the frequency. Here, genes P and Q will split up 35% of the time, giving them a "chromosomal distance" of 35 "map units." Genes R and Q will split up 20% of the time, giving them a chromosomal distance of 20 map units. The distances for the other genes can be read off the chart. Once you know all the distances, you can work out the relative positions of the genes to one another much like a puzzle. How can you fit together P, Q, R, and T in a linear pattern that matches the units in the chart? The answer is choice (C).

51. D Only choice (D) here is a Lamarckian explanation of why a vestigial organ would disappear. All other explanations are Darwinian in nature and can be considered valid reasons that a nonfunctional organ may be reduced in size to the point of disappearing. In (E), a negative genetic correlation between two traits means that the development of one trait is inversely proportional to the development of another. As selection favors sense organs other than eyes in cave-dwellers, eyes may disappear if the development of eyes were negatively correlated with the development of other sensory appendages.

52. C In an iteroparous life history, organisms save their energy for later reproductive efforts, delaying their maturity and investing energy early on in their own growth and self-maintenance. This type of reproductive strategy is found in species that live in environments where adult mortality is low and where the costs of reproduction are high, such that the organisms need to be large and strong before reproducing yet the reproductive effort is so great that they will be unlikely to repeat it (think of salmon spawning as adults only once at the end of their lives). In contrast, semelparous organisms reproduce once early on in life and then usually die. This kind of strategy is found where juvenile survival is high but the chances of making it to adulthood are low. It generally allows for extremely rapid population growth because the cycling of generations takes place much faster than in iteroparous species.

53. A Polygamy is a mating system that involves one male and many females. The most fit males are able to mate with a multitude of females, thereby passing their genes to a maximum number of offspring.

54. C Paedomorphies are features present in adults that are typical of features in the juvenile stage of that organism's ancestor. Paedomorphic features often arise because of a delay in growth of a particular organ, so that it remains underdeveloped in the adult. An example occurs in some salamander species whose adults retain the gills that are present only in juvenile salamanders of related species.

55. A Turner syndrome, otherwise known as monosomy X, occurs when a fertilized egg possesses only one X chromosome because of a nondisjunction event during male or female meiosis. Klinefelter syndrome occurs in males with genotype 47, XXY (they possess an extra X chromosome in every cell). Down syndrome is also known as trisomy-21, where individuals have three copies of chromosome 21 in every cell; and cri-du-chat syndrome occurs from a deletion of part of the short arm of chromosome 5. It was given its name because a crying infant affected by this syndrome sounds like a mewing cat.

56. D All of the statements are correct with the exception of choice (D). Estrogen is released within the ovaries, but acts anywhere in the body that estrogen receptors exist. This includes acting on cells in the hypothalamus and pituitary in order to regulate the secretion of GnRH, LH, and FSH.

57. D When fibers of a muscle are exposed to very frequent stimuli, the muscle cannot fully relax. The contractions begin to combine, becoming stronger and more prolonged. This is known as frequency summation. If the stimuli become so frequent that the muscle cannot relax, the contractions become continuous, which is known as tetanus.

58. E All of the answer choices should stand out as examples of why imperfections in the fossil record prevent paleontologists from readily identifying evolutionary patterns, except choice (E). If fossils show rapid transition from one phenotype to another over a short period, that may be indicative of punctuated equilibrium as a pattern for this species.

59. C Secondary growth results in the outward growth of a tree (in width, not height). During secondary growth, secondary xylem and phloem are built from secondary (vascular) cambium, and these tissues expand outward creating the familiar tree rings that can be studied to tell the age of a tree. The initial xylem and phloem of a tree is built from the embryonic tissue called procambium, which builds a cylinder of vascular tissue called the stele in young trees. The pericycle is the outer region of this stele.

60. C Lenticels, common in woody stems, are an avenue of gas exchange for very active tissues within the bark. Both the vascular cambium and the cork cambium are extremely active during growth and oxygen is needed for proper phloem transport as well. Pneumatophores, otherwise known as air roots, are present in trees such as mangroves, whose roots must use these extensions to exchange air above the water that submerges the base of the roots. Root hairs themselves exchange oxygen and carbon dioxide with small pockets of air within the soil, which is why you can kill a plant by overwatering it, thereby filling in these air pockets with water.

61. B Calcium ions are stored in muscle cells within the sarcoplasmic reticulum (SR), and they are pulled back into the SR after muscle contraction for use in the next contraction. All other answer choices deal with structures that are involved in the neuronal stimulation of the muscle fiber (using acetylcholine), not in the contraction itself.

62. E Plant X must be a C4 plant. The enzyme used by most plants (C3 plants) to capture carbon dioxide for the Krebs cycle is called rubisco. Yet rubisco tends to accept oxygen rather than carbon dioxide when oxygen concentrations are high. This process, known as photorespiration, consumes oxygen and generates no ATP. However, C4 plants, known for living in hot and dry climates, use the enzyme PEP carboxylase, which has a much higher affinity for carbon dioxide than rubisco. This allows these plants to continue accepting carbon dioxide, and hence photosynthesizing, long after oxygen concentrations rise around them.

63. A Light hitting the photosystems in the thylakoid membranes does provide the energy to produce ATP (a process called photophosphorylation, since the ATP is regenerated from ADP + P_i in the presence of light). Yet, this only indirectly involves the actual photosystem. The photosystems themselves are directly involved with processes such as the splitting of water to fill electron holes, receiving the light energy from the sun, and releasing oxygen.

In addition, chlorophyll a is present in both photosystems. Thus, the best answer here is that photophosphorylation is not directly associated with photosystem II.

64. C The fact that some modern-day bacteria live outside eukaryotic cells tell nothing of the origin of eukaryotic organelles such as the mitochondria and chloroplasts. All other statements listed accurately describe examples that support the endosymbiont theory, which proposes that many eukaryotic organelles were once their own free-living prokaryotic species.

65. C You learned back in the chapter on enzyme inhibition that folic acid is needed for nucleic acid biosynthesis. This vitamin is found in vegetables and whole grains and is also used for a variety of metabolic reactions involving amino acids. Vitamin B_6 is found in meats and vegetables and is needed for amino acid metabolism, while niacin is part of the coenzymes NAD^+ and $NADP^+$. Vitamin A is essential for healthy vision, and ascorbic acid (vitamin C) is used in collagen synthesis and as an antioxidant to combat damage due to oxygen radicals formed in certain cellular reactions.

66. E In angiosperms, embryos within the seed form one (monocot) or two (dicot) seed leaves that absorb endosperm and start the growth of the new plant. The single cotyledon of a monocot is also known as the scutellum.

67. D The pericarp is the thickened wall that surrounds the fruit of an angiosperm. The fruit begins to develop only after certain chemical changes that cause the ovary of the fruit to grow rapidly. As the fruit grows and matures, much of the rest of the flower that surrounds it deteriorates, eventually causing the fruit to fall to the ground.

68. B The gametophyte is the form of the plant that produces gametes via mitosis, in contrast to the spore-producing sporophyte. These haploid gametes will fertilize each other to develop into a single sporophyte organism.